THE MATH TUTOR'S FRIEND

language of algebra

fractions

handy math tips

signed numbers

mental math "tricks"

quizzes

BOOK 1

PRE-ALGEBRA

by
Joe Cyr

AF800790

Cyr, Joseph

The Math Tutor's Friend / *Book 1* - [illegible]

Third Edition ISBN: 978-1-967039 [illegible]

Second Edition ISBN: 979-8-9879 [illegible]

First Edition ISBN: 979845199952 [illegible]

3 4 5 6 7 8 9

Subjects: Main category of book – [illegible] *decimals; mental math; functions; algebra terms; number* [illegible]

2.Other category - math tutor refer [illegible]

Cover illustrations by [illegible] esign by Joe Cyr

B [illegible]

The [illegible] ***end***

Books

1 - Pre- [illegible]
2 - Alge [illegible]
3 - Alge [illegible]
4 - Alge [illegible]
5 - Prob [illegible] **s**
6 - Geo [illegible]
7 - Trig [illegible]
8 - A Lo [illegible]
9- Seri [illegible]
& " [illegible] **ndout**
[illegible]

B [illegible]
con [illegible] **s**
with [illegible] **d**

THE MATH TUTOR'S FRIEND

BOOK 1

Pre-Algebra

About the "Math Tutor's Friend" Series:

Teachers impart knowledge guided by lesson plans in a course syllabus. Math tutors, on the other hand, face varying topics at differing levels in successive sessions; for example, pre-algebra, advanced algebra, geometry, trigonometry, statistics, or even calculus! This series addresses the issue and is intended to provide the tutor with needed quick review of high-school math topics. The material is not presented rigorously nor does it abound with proofs and explanations that are available in textbooks - the tutor already knows this stuff - but a quick refresher before diving into a tutoring session is always a good idea.

Too many in high-school have reached the 11th and 12th grades lacking basic math skills such as memorizing the multiplication table (having to use a calculator to multiply 6 by 7, for example), handling signed numbers, and performing operations involving fractions - knowledge that should have been acquired by the fifth grade! I try to address this deficiency by devoting the first few tutoring sessions to these fundamentals and providing handouts of multiplication tables, etc.

Books 1 through 8 of this series include a section containing quiz questions (with answers explained) that the math tutor may find useful.

Book 9 is an overall index of topics. It contains a set of summary "FLASHMATH" pages for quick reference or for use as handouts.

This third edition corrects typos found in previous editions and expands discussion of some of the topics.

Happy Tutoring!

Joe Cyr

Sarasota, FL

THE MATH TUTOR'S FRIEND
prime number
order of operations
It's easy !!
+ - × ÷ ≈ > <
conjugate
rational number
variable
Mr. Math
FUN-DAMENTALS

About Book 1: "Pre-Algebr [illegible]

Section 1, *Fundamentals and T* [illegible] lication table, symbology, the number line, mathematical [illegible] tation, other fundamentals needed in the pursuit of upcom [illegible]

Emphasis -:

- the multiplication tabl [illegible]
- a fraction represents d [illegible]
- prime numbers
- observations relating [illegible]

Section 2, *Signed Numbers*, and [illegible] *d Decimals*, are especially important to the understanding [illegible] ematics.

Emphasis -:

- the operation "$a - b$" [illegible]
- least common denomi [illegible]
- how to rationalize a fr [illegible]
- "invert and multiply" [illegible]
- all numbers can be re [illegible]
- all decimals are fracti [illegible]

Section 4, *Handy Math Tips*, [illegible] s, shortcuts and presents handy techniques for mentall [illegible] mathematical operations. These probably should not be [illegible] t may be provided to the alert student who may wish to [illegible]

Emphasis -:

- dimension analysis
- divisibility rules

Section 5, *The Language of Al* [illegible] l jump-start into algebra

Emphasis -:

- "like" terms
- coefficients
- how to factor an algeb [illegible]

1. FUNDAMENTALS & TERMINOLOGY

TABLE OF CONTENTS

Continues . . .

TABLE OF CONTENTS (CONT'D)

•

1. FUNDAMENTALS & TERMINOLOGY

Symbols

Common math symbols

The common math symbols* include the *operators* +, –, ×, ÷ and the inequality symbols ≠, >, <, ≥, ≤, <<, >>. Page 5-9 of this book contains a table of these and other math symbols, their meanings, and the < Alt > codes to generate symbols not usually available on a standard computer keyboard.

Infinity

The symbol ∞ is used to represent "infinity." A *countable infinity* has no limit (*i.e.*, largest number). It is so-called because if you give me a number, say, and claim it to be the largest, I simply add "1" to it to make it larger. There are different levels of infinity and the concept itself is paradoxical. For example, the one-dimensional line has an infinity of points, but between any two given points, no matter how close, there is an infinity of points (a point has zero dimension).

Mathematical operators

The *mathematical operators* +, –, × (or "·"), ÷ (or "/" or ":") are addition, subtraction, multiplication, and division, respectively. These operators separate the terms in algebraic expressions. The multiplication operator is frequently implied in algebraic expressions when no confusion exists; for example, the term $3xy(2 + x)\cos(\theta)$ means 3 time x times y times $(2 + x)$ times $\cos(\theta)$.

Example: $3x^2 - 4(x + 3)/5$ contains the four mathematical operations.

There are several symbols or arrangements for product and division. The operator " · " is sometime used to indicate product, *e.g.*, $5 \cdot 7 = 35$; operator symbols for product are not needed when there is no confusion and the meaning is clear, *e.g.*, $(5)(7) = 35$, $5x$ is the product of 5 (called the *coefficient*) and the variable x. "xy" is the product of the variables x and y. In the case of division the *vinculum* or *bar* that separates the numerator and denominator in a fraction (and separates the dots in " ÷ "), is also called the *fraction bar*, and represents division, as does the *solidus* " / " symbol, *e.g.*, $(x + 5)/(2 + y)$ is $(x + 5)$ divided by $(2 + y)$. In fact, a fraction ***always*** represents division of the numerator by the denominator.

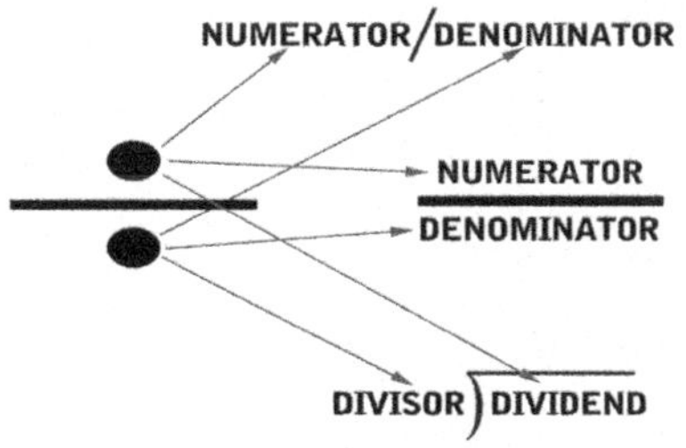

* Some symbols, terms and procedures in this section precede their introduction further on; but this series is intended for tutors who wish only to review the material.

The *ratio* 5:2 represents division [illegible] s used in discussions of proportion such as “five is to [illegible] f course, is 25 and is the solution to the *proportion* equ[illegible]

Variables.

Unknown quantities in math are rep[illegible] ols, or alphabet letters from other languages. In some cas[illegible] re assigned to certain characters.

Example: π = [illegible]

Subscripted variables

In simple problems, variables such [illegible] sufficient to represent unknown quantities; however man[illegible] s involve hundreds of variables. Subscripts provide a wa[illegible] ng variables with the same letter, say.

Example: x_3 differs from ever[illegible] in the sequence

[illegible]

Sigma notation to express a sum

The Greek letter capital *sigma* (Σ[illegible] subscripted terms or variables. The lower-case sigma [illegible] statistics.

Example: $\sum_{i=1}^{25} x_i = 256$ represen[illegible] ariables, x_1 through x_{25}.

The sym[illegible] *mation.*

Order of operations

The rules for the order of operati[illegible]

- 1) Do all operations within paren[illegible] enominators of fractions serve as parentheses). If there are nes[illegible] ([ ] [( )])],” work from inside out. (The *order of operations* ru[illegible] heses.)
- 2) Do any work with exponents or r[illegible]
- 3) Proceeding from left to right, do [illegible] isions.
- 4) Finally, working from left to righ[illegible] actions.

Before proceeding with Step (4), [illegible] many terms, it may be easier to group terms having the [illegible] ive numbers to get a positive sum, then add all of the [illegible] a negative sum. The result will be one algebraic addit[illegible] of two terms.

NOTE: Avoid using the divisio[illegible] ge it into a fraction. Additional rules apply when var[illegible]

Avoid confusion by using parentheses to group operations.

Example:

$$4(x + 2^3) + 26 - 12/3 - 25 + 1 - 3 = ?$$

$4(x + 8) + 26 - 12/3 - 25 + 1 - 3$ [Do the exponent inside the parentheses]

$= 4x + 32 + 26 - 12/3 - 25 + 1 - 3$ [Clear the parentheses]

$= 4x + 32 + 26 - 4 - 25 + 1 - 3$ [Do the division]

Now "collect" the positive terms and negative terms (there is only one "x" term)

$4x + (32 + 26 + 1) - (4 + 25 + 3)$ [Add within the parentheses]

$= 4x + 59 - 32$ [Add algebraically (*i.e.*, subtract)]

$= 4x + 27$ [the answer]

Types of Numbers

Real numbers

Informally, a real number is one that can be found on the *number line*.

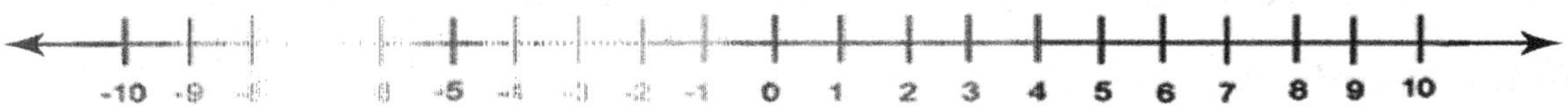

Natural numbers

1, 2, 3, … The "counting" numbers; you can count them on your fingers (up to 10, anyway).

Whole numbers

The natural numbers that include zero (a "hole"): 0, 1, 2, 3, ….

Integers

Whole numbers that may be positive or negative or zero: … -3, -2, -1, 0, 1, 2, ….

Integers are subdivided into *negative integers*, zero, and *positive integers* (does not include zero). Thus integers may be positive, negative, or zero.

Non-negative integers and non-positive integers include zero.

Positions of digits

Any integer will have a *units* (or *ones*) digit, and may have a *tens* digit, or *hundreds* digit, and so on.

Example: the number 4 has only a *units* digit because 4 × 1 = 4;
the number 35 has *units* digit 5 and *tens* digit 3 because (3 × 10) + (5 × 1) = 35;
the number 239 has *units* digit 9, *tens* digit 3, and *hundreds* digit 2 because (2 × 100) + (3 × 10) + (9 × 1) = 239. Get the idea?

Rational numbers

Numbers that can be expressed as the ratio of two integers.

Examples: ½, ¾, 3 (3 can be expressed as ³⁄₁).

Irrational numbers

Numbers that cannot be expressed as a ratio of two integers p and q – where p/q and $q \neq 0$. Examples: π, √2, $\log_2 3$.

Transcendental numbers

A number (which may may be complex) that is not a root (*i.e.*, solution) of a nonzero polynomial equation with integer coefficients.

Examples: π, e, ***i***π.

Odd and Even numbers

The integers -2, 2, 4, 6 are even. If an integer is neither even nor zero then it is ***odd***, *e.g.,* -5, -3, 3, 5, 7. Odd and even integers alternate; that is, if integer m is even, then the integers $m - 1$ and $m + 1$ are odd. If you want to specify an arbitrary odd integer, take any integer n, multiply it by 2 to make it even, then add or subtract one to the result: $2n$ is always even, so $2n + 1$ and $2n - 1$ are always odd. Decimals (such as 4.3 and 3.24) are neither odd nor even.

Prime numbers

Prime numbers are positive integers divisible only by "1" and themselves. The first few *prime numbers* are 2, 3, 5, 7, 11, 13. Here are the prime numbers less than 500:

2, 3, 5, 7, 11, 13, 17, 19, 23, 29, 31, 37, 41, 43, 47, 53, 59, 61, 67, 71, 73, 79, 83, 89, 97, 101, 103, 107, 109, 113, 127, 131, 137, 139, 149, 151, 157, 163, 167, 173, 179, 181, 191, 193, 197, 199, 211, 223, 227, 229, 233, 239, 241, 251, 257, 263, 269, 271, 277, 281, 283, 293, 307, 311, 313, 317, 331, 337, 347, 349, 353, 359, 367, 373, 379, 383, 389, 397, 401, 409, 419, 421, 431, 433, 439, 443, 449, 457, 461, 463, 467, 479, 487, 491, 499.

Look at the list of primes; note that, except for "2," they all end in (*i.e.*, the *units* digit is) 1, 3, 7, or 9. Why is that?

When is a number not a prime?

Given a number, can you positively state that it is NOT prime? For example, the integer 34,569 is not a prime number because it is divisible* by 3.

If the "ones" digit is a 0, 2, 4, 6, or 8, then the number is even (divisible by 2) and (except, of course, 2) cannot be prime.

If the "ones" digit is 5, then the number is divisible by 5 and so cannot be prime.

Therefore, if a number is prime, then its "ones" digit must be 1, 3, 7, or 9.

Coprime

(also relatively prime)

Numbers (not necessarily prime) that have no common factors are coprime.

Example: 11 and 12 are coprime even though 12 is not prime.

* See the divisibility rules in Section 4.

Opposite

When the *opposite* (of a number) is added to the number, the sum is 0.

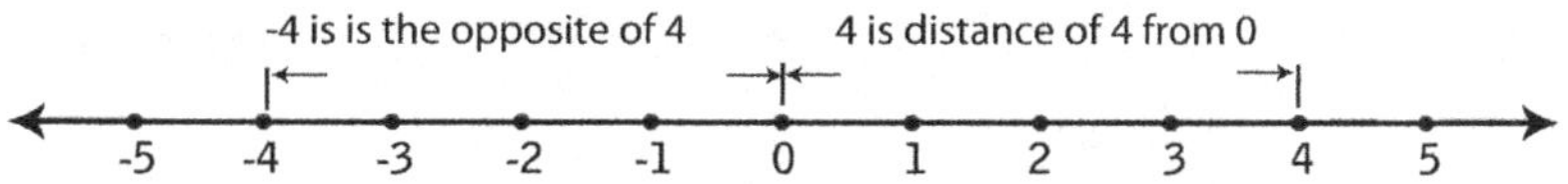

A number and its opposite are the same distance from 0

Example: The opposite of x is $-x$; the opposite of 4 is -4.

Complex numbers

Imaginary number

An *imaginary number* is a *complex number* that can be written as a real number multiplied by the imaginary unit $\boldsymbol{i}$ (defined by its property $\boldsymbol{i}^2 = -1$). The square of an imaginary number $\boldsymbol{bi}$ is $-b^2$. An imaginary number $\boldsymbol{bi}$ can be added to a real number a to form a complex number of the form $a + \boldsymbol{bi}$. The real numbers a and b are the *real part* and the *imaginary part,* respectively, of the complex number. Sometime the notation used is $\boldsymbol{ib}$, or even $\boldsymbol{jb}$ (usually in engineering).

Conjugate

Just change the signs in a binomial expression to obtain its conjugate.

Example: $a - b$ is the conjugate of $a + b$ and vice-versa.

Multiplying a binomial by its conjugate results in the *difference of squares*, a pattern that should be memorized; a difference of squares in an algebraic expression can be replaced by the product of a binomial and its conjugate and vice-versa.

Example: $(a - b)(a + b) = a^2 - b^2$

Properties of real numbers

All real numbers, say x, y, and z, have the following properties:

Symmetric: If $x = y$ then $y = x$

Transitive: If $x = y$ and $y = z$ then $x = z$

Example: If $x = y$ and $y = 9$ then $x = 9$.

Substitution: If $x = y$ then x may replace y in any equation or expression and vice-versa. Here, x and y may be variables, terms, or expressions

Reflexive: $x = x$

Examples: $3 = 3$; $0 = 0$

Commutative: A change in the order of the addends does not change the sum.

Example: $x + y + z = y + z + x$

A change in the order of the numbers to be multiplied does not change the product.

Example: $x \times y = y \times x$ ($xy = yx$; $x{\cdot}y = y{\cdot}x$)

Associative: The sum or product of [illegible] terms, or expressions, remains the same regardless of how [illegible]

Examples: (Addition) 3 [illegible] 5) + $(x + y)$ + 6
(Multiplication

Distributive: $x(y + z) = xy + xz$. [illegible] $\pm\ xz$

Similarly, given $x, y, z \neq 0$, $x(y \div$ [illegible]

Reciprocal

The *reciprocal* of a number or exp[illegible] or expression divided into "1."

Examples: 1/3 is the recip[illegible] (1 ÷ 1/3 = 3)
$1/(5x^2 + 4)$ is the recip[illegible] [illegible]ce-versa.

Identity

An *identity element* (or *identity va*[illegible] leaves unchanged any expression in which it occurs.*

Multiplicative identity: $x \times$ [illegible] y number and its multiplicative identity valu[illegible] as the result).

Additive identity: $x + y + 0$ [illegible] number and its additive identity value giv[illegible] the result).

Inverse

Additive inverse - A numb[illegible] sums to zero
Examples: [illegible]

Multiplicative inverse - A [illegible] reciprocal equals 1.
Examples: 15 × ([illegible] = 1.

Observations relating to 0

The number "0" is not necessari[illegible] ple, in math "0" may be thought of as separating the p[illegible] rom the negative real numbers; in the number 201, "([illegible] t separates the units digit from the hundreds digit; in t[illegible] first zero indicates the absence of an integer part, and t[illegible] place-holders for the tenths and hundredths digits in th[illegible] umber; the "0" in the number 5.1230 indicates that the [illegible] four decimal places.

Examples: number li[illegible] 0.03060.

Rules involving zero

- $0 \times x = 0$, where x is any numb[illegible] xpression.
- $0 \div x = 0$, where x is any numbe[illegible] ression and $x \neq 0$.

* Any number or expression divided by an id[illegible] $(3x + 34)/1 = (3x + 34)$.

• $x^0 = 1$, where [illegible] is any number, variable, term or expression.*
• 0/0 (or 0 ÷ 0) [illegible] defined; $x/0$ is undefined; y/x is undefined if $x = 0$; the rational function [illegible] $G(x)$ is undefined at values of x where $G(x) = 0$.

Observation[illegible] [illegible]ting to 1

• $1 = x^0 = 1 \cdot 1$ [illegible] $n^2\theta - \cos^2\theta) = -(-1) = x/x =$ [anything]^0

• $x = 1x = x/1$ [illegible] $= x + 0 = -(-x) = -1(-x)$

Knowing th[illegible] [illegible]ve is useful when dealing with fractions, exponents, identities, *etc.*

The multipli[illegible] table

Many student[illegible] [illegible] a calculator to determine, say, 6 times 7. The multiplication table on page [illegible] should be committed to memory by the fifth grade! I always provide a copy [illegible] table to each student with the advice to memorize it.

Interpolation [illegible] estimating a value between two known values, say on a graph or a me[illegible] stick or a pressure gauge, *etc.* See page 1-10.

Reading and [illegible]essing numbers

As a tutor you [illegible] need to provide guidance in the correct reading of numbers.

Example: Th[illegible] [illegible]er 64,520,264.03529 reads as "sixty-four million five hundred twenty [illegible] [illegible]d two hundred sixty-four and three thousand five hundred twenty-nine [illegible] thousandths."

As important [illegible] ability to read the entire number is the ability to identify each place to [illegible] [illegible] and to the left of the decimal point; one may speak of "the thousands dig[illegible] [illegible] the "hundredths digit," for example. See the figures below.

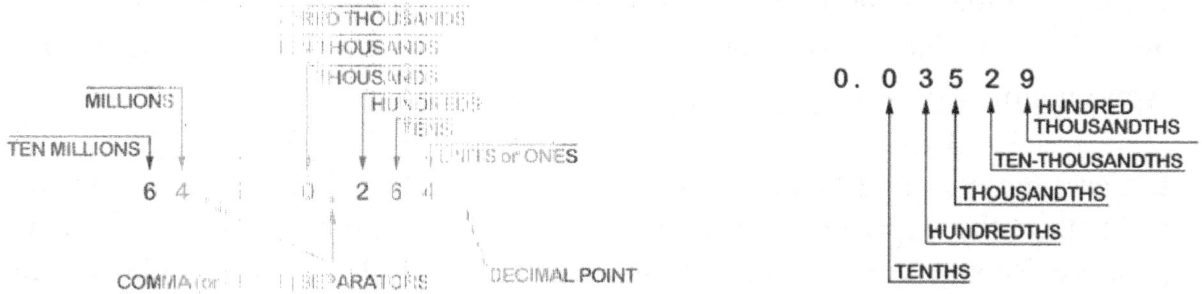

Notice the co[illegible] [illegible] regarding the separators. While we in the United States (and Liberia) [illegible] write 64,520,264 the rest of the world, which uses the metric system, w[illegible] [illegible] spaces (instead of commas) as separators and write it as 64 520 264. I[illegible] [illegible] metric system; 4-digit metric numbers are shown without commas (*e.g.*, [illegible] m); five or more digits use spaces instead of commas (*e.g.*, 3 344 000). In [illegible] countries 64,520,264 would be written as 64.520.264; and, yes, the decim[illegible] [illegible]" would be written as "3,4" (the comma is called *virgule*).

* 0^0 is sometimes u[illegible]d.

** Compare with [illegible]*ation*: the extension of a range of values by inferring unknown values from perceived trends in [illegible]n data.

Math prefixes

Number prefixes such as *mega, kilo, centi,* and *deci* are common. The table on the next page summarizes the math prefixes along with the corresponding scientific notation. See page I-22 of this section for a more comprehensive table of prefixes.

Scientific notation

Engineers and scientists handle very large and very small numbers by using scientific notation. For example, multiplying 0.00000021 by 3,450,000,000 is not easy to do by hand and is awkward using a calculator. So this is how it's done (negative numbers are handled the same way): For any number, move the decimal place right or left as necessary to produce a number between 1 and 9 to the left of the decimal point (*i.e.*, 1., 2., 3., 4., 5., 6., 7., 8., 9.).

Example: 0.00000021 becomes 2.1 by moving the decimal of 0.00000021 to the right seven places and 3,450,000,000 becomes 3.45 by moving the decimal point nine places to the left. Well, doing this certainly changes the number so we must account for this change by multiplying the modified number (2.1 and 3.45 in our example) by "10" raised to a power determined by the number of places moved: a negative integer if movement was to the right and a positive integer if the movement was to the left. Note that $5 = 5 \times 10^{0}$ in scientific notation.

Examples: $0.00000021 = 2.1 \times 10^{-7}$

$3{,}450{,}000{,}000 = 3.45 \times 10^{9}$.

To recap this important procedure: for the example, change 0.00000021 to 2.1, by moving the decimal point to the right seven places, and write 2.1×10^{-7}. Change 3,450,000,000 to 3.45 by moving the decimal point left nine places to get 3.45×10^{9}. Now, $0.00000021 \times 3{,}450{,}000{,}000 = 2.1 \times 10^{-7} \times 3.45 \times 10^{9}$

Applying the *Law of Exponents*, the problem becomes

$$2.1 \times 3.45 \times 10^{9-7} = 2.1 \times 3.45 \times 10^{2} = 7.245 \times 10^{2} = 724.5.$$

It's as simple as that! The rule is that if you move the decimal point to the right, then the exponent is negative; move to the left then the exponent is positive. Be sure you understand this from the example above. Numbers with many digits are handled by a procedure of rounding off to a specified number of significant figures so that one can produce an answer to the required degree of accuracy. If you wish to pursue this further, use a search engine like Google to look for "significant figures" and "rounding numbers."

Example: Multiply 0.12345678 by 98,234,367,899

The previous rule seems to be of little help. We must resort to the real-world tactic of rounding. Let's round to three decimal places after moving the decimal point:

$$0.12345678 \approx 1.235\times10^{-1} \text{ and } 98{,}234{,}367{,}899 \approx 9.823\times10^{10}.$$

The symbol " ≈ " means "approximately." The above example employed both rounding and scientific notation to arrive at an approximate solution. Ideally the numbers should be multiplied first and then the product rounded. The table below lists numerical prefixes for numbers.

MATH PREFIXES				
PREFIX	SYMBOL	1 ×	WORDS	EXAMPLE
Tera	T	10^{12}	one trillion	terabyte
Giga	G	10^{9}	one billion	gigahertz
Mega	M	10^{6}	one million	megawatt
Kilo	k	10^{3}	one thousand	kilometer
Hecto	h	10^{2}	one hundred	[seldom used]
Deka	da	10^{1}	ten	[seldom used]
unit		10^{0}		
Deci	d	10^{-1}	one tenth	decibel
Centi	c	10^{-2}	one hundredth	centimeter
Milli	m	10^{-3}	one thousandth	milliliter
Micro	µ	10^{-6}	one millionth	microsecond
Nano	*n*	10^{-9}	one billionth	nanosecond
Pico	p	10^{-12}	one trillionth	picofarad

Accuracy and rounding

Measurement

If you want to know a physical attribute such as length, area, volume, weight, mass, angle, temperature, pressure, *etc*., then you will take a measurement to estimate the quantity of units of that attribute. A *measurement* is an estimate to a specified degree of accuracy determined by the device or instrument that is used.

Significant figures / significan digits / significant numbers

Significant figures (*sig figs*) apply only to measured values. The last (*i.e*., right-most) digit in a measurement is an estimate that may be referred to as the *doubtful digit* or *estimated digit*. If you are measuring an object with a ruler that has ten marks between whole centimeters, for example, then it makes no sense to report a length as 3.26278 cm because we can only estimate measurements between the hash marks on the ruler. See the figure below: the measurement is between 3.2 and 3.3 with a

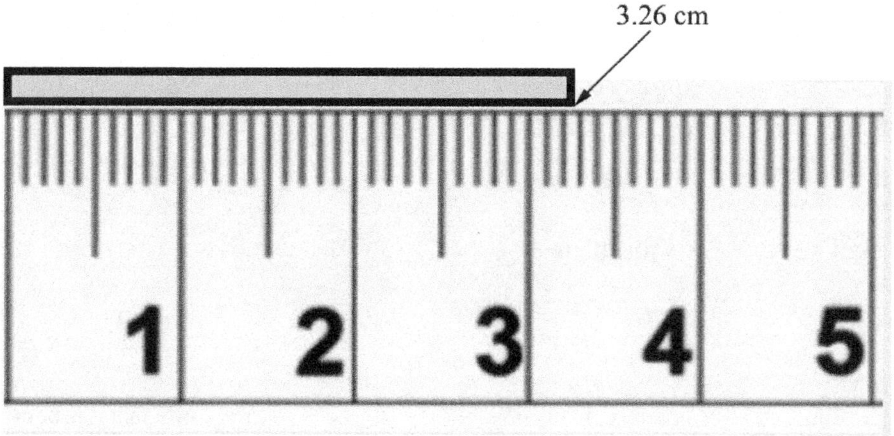

guess at the hundredth place (here, [illegible]). Only [illegible] digits as allowed by the *measurement resolution** are reliabl[illegible] the significant figures.

The last digit is an estimate

The last digit in a measuremen[illegible]e. For example, the measurement of 3.26 cm shown [illegible] the length was a little more than half-way between 3.[illegible] "6" in the hundredth place, then, is a guess but is still [illegible]

Examples of significant digits

- 1000 has one significant digit: only [illegible]cific because the zeroes are simply placeholders and could [illegible]nding to get the value 1000.
- 1000.0 has five significant digits: [illegible] the decimal point is the presumed accuracy of the measuren[illegible]surement is accurate to the tenths place, in this example, ze[illegible]
- 0.00083 has two significant digits: [illegible] are placeholders only providing information about relative [illegible]
- 0.000830 has three significant digi[illegible] that the measurement is accurate to that last digit, which is ze[illegible]
- 2004 has four significant digits: the [illegible] and we have to count the zeroes because they are between the [illegible]
- 210 has two significant digits: the las[illegible] in the units place.
- 210. Note the decimal point after the [illegible] number has three significant digits. The decimal point signifies that [illegible]de to the nearest unit, so the zero is not just a placeholder.
- 470.0 has four significant digits: the [illegible]ans that the measurement was made accurate to the tenths place, [illegible]; the 4 and 7 give useful information, and the units place-holde[illegible]ant digits, ("7" and ".0") and so is also significant.

Determine the significant digits [illegible]

The basic rules are

All non-zero d[illegible]its.

For example,

- 6 has one significant digit;
- 3.2 has two significant digits;
- 4,193.753 has seven significant di[illegible]

* Measurement resolution is the smallest chan[illegible]ntity that it is being measured.

Zeros between significant digits are significant digits.*

For example,

- 301 has three significant digits
- 3.001 has four significant digits
- 30.001 has five significant digits
- 30000.0 has six significant digits

Zeros to the right of the decimal point
that are also to the right of a non-zero digit are significant.

For example,

- 0.050 has two significant digits (the zero to the left of the 5 is not significant)
- 0.0020 has two significant digits
- 10.40 has four significant digits
- 710. has three significant digits
  BUT "710" has no decimal points, so has two significant digits.
- 25.4000 has six significant digits
- 100.00 has five significant digits (The two zeros to the right of the decimal point are significant. The two zeros to the left of the decimal point are significant because they lie between significant digits, *i.e.*, the "1" and the significant zeros to the right of the decimal point.)
- $54\bar{0}0$ has three significant figures (indicated by the overbar on the first zero).

A zero to the right of the decimal point
must also be to the right of a non-zero digit to be significant.

Examples: 0.00024 has two significant figures
0.000240 has three significant figures.

Determine the number of significant digits - *Mnemonic:*

On a standard map of the U.S. the Pacific Ocean is to the Left of the continent and the Atlantic is to the Right. If a number includes a decimal point, then the decimal is Present (*Pacific*), count significant numbers from the Left; if a decimal point is Absent (*Atlantic*), then count significant numbers from the Right.

Examples:

- 3400. the decimal point is Present: count 4 significant figures from left to right.
- 3400 the decimal point is Absent: count 2 significant figures from right to left.
- 420.0001300 the decimal point is Present: count 10 significant figures from left to right.
- 0.0001300 the decimal point is Present: count 4 significant figures from left to right.
- 0400230 the decimal point is Absent: count 5 significant figures from right to left.

* These are also called *trapped zeroes* or *embedded zeroes.*

Significant figures in scientific notation

Numbers preceding "$\times 10^n$" in scientific notation (*i.e.*, the *mantissa*) are significant.

Examples:

- 1×10^{12} has one significant digit;
- 0.03×10^{10} has two significant digits;*
- 1.02×10^{8} has three significant digits;
- 2.700×10^{-4} has four significant digits.

Computations involving significant numbers

Important: measurement precision cannot be improved by mathematical operations.

Addition and subtraction

Adding measurements

With respect to measurements, adding a number with an estimated digit in the tenths place to another number with an estimated digit in the hundredths place will yield a sum with doubt in the tenths place. Numbers after the tenths place are meaningless. Answers may have but one estimated digit and no meaningless digits.

By hand or on a calculator: Note that 2.5 + 31.31 + 10.025 – 1.1 sum to 42.735; however the sum should be reported as 42.7 because the decimal ".735" is the sum of estimated digits in the tenth, hundredth, and thousandth decimal places, the least of which is the tenth, and so there should be only a tenth estimated decimal in the answer (".7"). See the hand calculation for this problem on the left; here, each estimated digit is circled. All three digits to the right of the decimal point are estimates in the sum because each is the result of adding at least one estimated digit. There must be only one estimated digit in the answer so the correct sum is 42.7, rounded to the tenths place.

```
   2 .(5)
  31 . 3(1)
  10 . 0 2(5)
  -1 .(1)
 ------------
  42 .(7)(3)(5)
```

Another way to look at addition with estimated digits

Let's say that one object is measured to have a mass of 8.7 gm (recall, the ".7" is an estimate) and a second object is measured on a different balance to have a mass of 0.3253 gm ("3" in the right-most [ten-thousandths] place is the doubtful digit). In adding these two measurements, write the numbers for five decimal places with "#" marks at digit positions where we lack information. Thus 8.7#### gm and 0.3163# gm. Adding ("#" + a digit = "#"), we have

```
8.7####
0.3163#
-------
9.0####
```

The sum is 9.0 gm. to two significant digits.

* If this were to be written in standard scientific notation, the *mantissa* would be a digit between 1 and 9; 3×10^8, however, would incorrectly show the value to have only one significant digit.

More comments on addition and subtraction

In a given number, the *most significant digit* is the digit with the highest exponent value (the left-most significant figure), and the *least significant digit* is the digit with the lowest exponent value (the right-most significant figure). For example, in the number "625.37", the "6" is the most significant figure as it counts hundreds (10^2), and "7" is the least significant figure as it counts one-hundredths (10^{-2}). The *least precise number* has the fewest digits to the right of the decimal point. The calculated sum or difference will have the same number of digits to the right of the decimal point as that of the *least* precise quantity of the numbers to be added or subtracted.

Examples:

(1) $100 - 1 = 99 = 100$ rounded up to one significant figure.

(2) $3 + 2.1 = 5$

The 3 is significant to the ones place, 2.1 is significant to the tenths place. Of the two, the least precise is 3 because it has the fewest (none) digits to the right of the decimal point. The answer cannot have any significant figures past the ones place.

(3) $6.0 + 3.1 + 2.2 = 11.3$

All the addends are significant to the tenths place, so the answer is significant to the tenth place. While each term has two digits of significance, the sum carried over into the tens columns so the answer has three digits of significance.

(4) $1{,}000 + 2{,}100 = 3{,}000$

The addend 1,000 has one significant digit and so must the answer have only one significant digit (remember, we are dealing with measurements here).

(5) $1{,}000. + 2{,}100. = 3{,}100.$

Note the decimal points. Both addends are significant to the *ones* place, and so the sum will also be significant to the ones place.

(6) $3.1 \times 10^2 + 212 = 5.2 \times 10^2$

The mantissa of the number in scientific notation is less precise than 212 (2.12×10^2) so the sum will have one decimal place.

When adding or subtracting measurements, find the left-most decimal place that contains an estimated digit. Round your answer to this decimal place.

Multiplication and division

Multiply measurements

In the case of multiplication we look for the significant digits of the multiplicand and the multiplier and express the product to the number of significant digits equal to the least of these. For division, the quotient will have significant digits equal to the lesser of the divisor and dividend.

Example: The product of 4.342 and [illegible] hand as

```
 4.342        four significan[illegible]
   2.8        two significant [illegible]
 34736
 8684
12.1576       must be rewritten to t[illegible]
              Therefore, the product is [illegible] 12
                        4.[illegible]
```

Example:

The calculator displays the produ[illegible] [illegible]our significant figures. However, you multiplied a number [illegible] [illegible]es by one having three significant figures. Your answer ca[illegible] [illegible] than the least-precise number in the problem. Therefore, t[illegible] [illegible] number as significant as the least-significant number to b[illegible] [illegible]ule applies to division (again, we're dealing with measure[illegible]

Summary

Addition and subtraction:

The number of decimal places i[illegible] [illegible] smallest number of decimal places in the input numb[illegible]

Multiplication and division;

Step 1: Among the numbers being [illegible] [illegible]ividend and divisor), identify the limiting term, the term [illegible] [illegible]r of significant digits.

Step 2: Multiply or divide as usu[illegible]

Step 3: Round the product (or q[illegible] [illegible]umber number of significant digits as the limiting [illegible]

Example: 7.1446 m × 4.8 m = 34.2[illegible] [illegible]wo significant digits.

Rounding numbers

A number is rounded off by dro[illegible] [illegible]nbers from the right and adding zeroes, if necessary, [illegible] [illegible]lative to the decimal point. When determining the n[illegible] [illegible]igures needed in the final answer, follow these rules:

Addition/subtraction: the l[illegible] [illegible]ace.

Multiplication/division: the [illegible] [illegible]ant digits.

Rounding numbers is important w[illegible] [illegible] significant figures in calculations. In everyday life, rou[illegible] [illegible]lating a tip or dividing a bar tab. There are many rules fo[illegible] [illegible]mbers are rounded to reduce the number of digits to the [illegible] [illegible]int.

Example: 3.141[illegible] [illegible]14.

Integers are r[illegible]d in order to simplify communication, a process done for convenience, [illegible]curacy. Example: 2,345,843 rounds down to 2,000,000 or "2 million," the l[illegible] term more likely to be used by news media. Computers and calculators pe[illegible]n intermediate calculations in the background that involve rounding. Diff[illegible] rounding-rules can produce vastly different end results, so a good practice [illegible] round only the final answer.

Types of roun[illegible]

- Rounding sp[illegible], by a teacher / employer / policy / *etc.*
- Odd and ev[illegible]unding
- Rounding m[illegible]ements (significant digits) [*exact values* are not rounded.]
- Rounding to [illegible]pecified number of decimal places
- Rounding in[illegible]s to units, tens, hundreds, *etc.* (government or media reporting)
- Rounding U[illegible]ollar amounts
    - nearest [illegible]r
    - roundi[illegible]own
    - roundi[illegible]p
- Truncating

Round to a s[illegible]ied number of decimal places:

Truncating d[illegible]al numbers:

Let x = 32.7[illegible], [illegible]n the function INT(x) = 32; the decimal is truncated or "chopped off.[illegible] positive numbers, the integer function INT(x) and the floor function yiel[illegible]ame results:* INT(32.76) = 32.

Round only [illegible]e final calculation

As already me[illegible]ed, in multiple-stage operations, do not round intermediate stage results; keep a[illegible]y digits as is practical until the end of all the calculations to reduce cumula[illegible]ounding errors. Round the final result to the least number of significant fig[illegible]ultiplication or division) and the left-most last significant digit position (addit[illegible] subtraction) among the inputs.

Rounding of [illegible]iplication and division revisited

Assume that t[illegible]mber "8" is a measurement (and therefore is estimated). The number "8" re[illegible]nts an inexact measurement with only one significant digit.; therefore, the r[illegible]of "8 × 8" is rounded to one significant digit, *i.e.*, "6 × 10 ¹". A measurement [illegible]" lies between 7.5 and 8.5 and so 8 × 8 ranges between 56.25 and 72.25. So [illegible]0¹ is the best one can give, as other possible answers give a false sense of [illegible]y.

* These functions [illegible]e integer less than or equal to the number so for negative numbers the integer is changed. For e[illegible] INT(-4.25) = -5

Examples:

- $8 \times 8 \approx 6 \times 10^{1}$ Answer is one significant digit
- $8 \times 8.0 \approx 6 \times 10^{1}$ Answer is one significant digit
- $8.0 \times 8.0 \approx 64$ Answer is two significant digits
- $8.02 \times 8.02 \approx 64.3$ Answer is three significant digits
- $8 / 2.0 \approx 4$ Answer is one significant digit
- $8.6 / 2.0012 \approx 4.3$ Answer is two significant digits
- $2 \times 0.8 \approx 2$ Answer is one significant digit

Rounding a decimal number to integer

The terms "round to the nearest hundred" and "round up to the nearest dollar" are familiar to most of us. In math, "round to the nearest integer" means convert decimal numbers to integers.

Example: "Rounding 35.67 to the nearest integer" would yield 36. This is like "round $35.67 to the nearest dollar."

Let's say you must round 42.5 to the nearest integer. Do you "round up" to 43 or "round down" to 42? There are various rules for rounding when the *right-adjacent digit* is 5.

Example: Round 43.5 and 74.5 to the nearest integer. Sometimes this is called "tie-breaking." One rule (there are many other rules) is that if the integer part of the decimal is odd, then round up to the next higher integer, otherwise *truncate* ("chop off") the 0.5; thus, 43.5 rounds up to 44. and 74.5 truncates to 74.

Rounding U.S. monetary values

Sometimes - in dealing with prices perhaps when it is inconvenient to handle coins - rounding is specified by policy: "round up to the next dollar" so $43.36 rounds up to $44; or "round down" means just drop the cents so $43.63 rounds down to $43. "Round to the nearest dollar" means that $43.36 rounds down to $43 and $23.52 rounds up to $24.

Rounding numbers involves the rounding digit.

When working with whole numbers and rounding to the closest 10, the *rounding digit* is the 10's digit (second number to the left of the decimal point). When rounding to the nearest hundred, round the 100's digit, and so forth.

First, determine what your rounding digit is and then look to the digit at its right (the *right-adjacent digit*):

• If the right adjacent digit is 0, 1, 2, 3, or 4, do not change the rounding digit. All digits to the right of the rounding digit become 0.

Example: Round 123 428 to the nearest thousand
The rounding digit is the thousands digit 3 and its right adjacent digit is 4
Thus 123 428 rounds to 123 000.

• If the right adjacent digit is 6, 7, or 8, increase the rounding digit it by one to the next higher digit. All digits to the right of the rounding digit become 0. If the rounding digit is 9, then change it to zero and increment the left-adjacent digit by one.

Example: Round 129 754 to the nearest thousand.
The rounding digit is the thousands digit 9 and its right-adjacent digit is 7
so 129 754 rounds to 130 000

Example: Round 398 838 to the nearest ten-thousand.
The rounding digit is the ten-thousands digit 9 and its right-adjacent digit is 8
so the 9 rounds to 10.
and 398 838 becomes 400 000.

• If the right-adjacent digit is 5 use the following "tie-breaking" rules:

• • At least one non-zero digit follows the 5, increase the rounding digit by one.

• • All digits to the right of the 5 are zero ***AND*** the rounding digit is 0 or even; replace the 5 with a zero.

• • All digits to the right of the 5 are zero ***AND*** the rounding digit is odd; replace the 5 with a zero and increase the rounding digit by one.

Example: Let's round the following numbers: 3,465,001, 3,465,000, and 3,475,000, using the above tie-breaking rules:
Round to nearest ten-thousand . . .

3,465,001	rounding digit is "6" ; round 3,465,001 to 3,470,000.
3,465,000	rounding digit is "6" ; round 3,465,000 to 3,460,000.
3,475,000	rounding digit is "7" ; round 3,475,000 to 3,480,000.

Rounding the Decimal-part of a Number
Identify the rounding digit and its right-adjacent digit:

• If the right-adjacent digit is 4, 3, 2, or 1, simply drop (truncate) all digits to the right of the rounding digit.

Example: 0.346247, rounding thousandths digit 6, rounds to 0.346

• If the right-adjacent digit is 6, 7, 8, or 9 increase the rounding digit by 1 and drop all digits to the right of it.

Example: 0.346247, rounding hundredths digit 4, rounds to 0.35

• If the right-adjacent digit is 5 and its adjacent digit to the right is 5, 6, 7, 8, or 9, increase the rounding digit by 1 and drop all digits to its right.

Example: 0.345677, rounding hundredths digit 4, rounds to 0.35.

• If the right-adjacent digit is 4 and its adjacent digit to the right is less than 5, then simply drop all digits to the right of the rounding digit.

Example: 0.213446, rounding thousandths digit 3, rounds to 0. 213.

There are other rounding rules such [illegible] is odd, then an adjacent "5" rounds it up, otherwise, drop al [illegible] ule for rounding up for "5" is that if its predecessor is an [illegible] to the next even digit, *e.g.*, 0.2375 rounds up to 0.238; oth [illegible] number to round "up" is even then drop the 5. Thus 0.238 [illegible] 345 rounds to 0.34. As you might guess, rounding can becc [illegible] ver rule you employ, be consistent.

Examples: 765.3682 becomes:

- 1,000 when rounding to the nea [illegible]
- 800 when rounding to the neare [illegible]
- 770 when rounding to the neare [illegible]
- 765 when rounding to the neare [illegible]
- 765.4 when rounding to the nea [illegible]
- 765.37 when rounding to the ne [illegible]
- 765.368 when rounding to the n [illegible]

[Much of the material above is from Russe [illegible] nding Numbers Correctly." ThoughtCo, Feb. 11, 2020, thoughtco.com [illegible] 079.]

Numbers presented to many decim [illegible] work with; the decimal 3.141592654 is the constant π to nir [illegible] e π is to be rounded to eight decimal places. We drop (trur [illegible] .14159265. Rounding to seven decimal places yields 3.141 [illegible] The required rounding is usually specified in the problem s [illegible] three decimal places: 3.142," or "round π to the fraction 2 [illegible] ed, rounding produces errors, so do rounding in the last ste [illegible].

Summary of significant digits

Significant figures apply only to me [illegible] own to be exact should be ignored for determining the numl [illegible] hat belong in the result. Examples of exact values include:

- integer counts (*e.g.*, the number o [illegible]
- definitions of one unit in terms of [illegible] 60 seconds, one inch is 2.54 centimeters);
- actual prices and costs;
- quantities given in requirement spec [illegible]
- legally defined conversions, such as [illegible] change;
- scalar operations, such as "doubling," [illegible]
- mathematical constants, such as π [illegible]
- quantities expressed in radicals su [illegible]
- Physical constants that were deter [illegible] ave significant digits.
  An exception is c (the speed o [illegible] be exactly 299,792,458 meters/second.

Combining exact numbers with estimated numbers
Values that are exact should be ignored when determining the number of significant figures that belong in the result. Exact numbers have an infinite number of significant figures, so never have the least number of significant figures. Combining exact numbers with estimated numbers is straightforward:

Addition/subtraction: accuracy is determined by the decimal places. Let 2.54 be an exact number and 3.141 an estimated number, then 2.54 + 3.141= 5.681.

Multiplication/division: accuracy is determined by the significant figures of the estimated number. Let 2.54 be an exact number and 3.142 an estimated number, then 2.54 × 3.142 = 7.97814 is rounded to four significant digits: 7.978

Effect of rounding
Rounding, addressed earlier in this section, is often employed for "ball-park" estimates and for convenience - a project costing $8.6 million, for example.

Do you suppose that rounding could possibly be exploited by "working the system?" For example, Congress could provide a contractor with a "$1.2 billion" contract and actually approve $1,243,999,999 (which rounds down to $1.2 billion), then actually pay $1,149,999,999 (which rounds up to $1.2 billion). The difference is $94 million dollars unaccounted and could be used for what? No doubt there are safeguards to prevent this. [illegible]?

Rounding to significant digits & to decimal places compared.
This all sounds complicated, and it is; there are many considerations and exceptions that researchers must know about, but we are just interested in techniques and making the reader aware of the many facets of significant figures and rounding and the handling of measurements.

MORE EXAMPLES:

Numbers of significant digits (sig figs).

Number	*sig figs*	*Scientific notation*
0.00682	3	6.82×10^{-3}

(Leading zeroes are not significant.)

Number	*sig figs*	*Scientific notation*
1.072	4	$1.072\ (\times 10^{0})$

(Embedded zeroes are always significant.)

Number	*sig figs*	*Scientific notation*
300	1	3×10^{2}
300.	3	3.00×10^{2}
300.0	4	3.000×10^{2}

(Trailing zeroes are significant only if the decimal point is specified.)

A final word about operations with significant figures (*i.e.*, measurements)

• The result is always determined by the *least precise number* (measurement) in the problem.

• Rules for algebraic addition differ from those for multiplication & division.

•• ***algebraic addition:*** deal only with the decimal portion (*i.e.*, digits to the right of the decimal point) of the numbers; ignore numbers to the left of the decimal point in determining precision.

1) Determine the least number of significant figures in the decimal portion of the numbers in the problem.

Round 62.567 to			Round 0.062567 to		
Significant digits	Precision	Decimal places	Significant digits	Precision	Decimal places
62.56700	7	62.5670000	0.06256700	7	0.0625670
62.5670	6	62.567000	0.0625670	6	0.062567
62.567	5	62.56700	0.062567	5	0.06257
62.57	4	62.5670	0.06257	4	0.0626
62.6	3	62.567	0.0626	3	0.063
62	2	62.57	0.063	2	0.06
60	1	62.6	0.06	1	0.1
-	0	63	-	0	0

2) Do the addition.

3) Round the answer to the number of decimals determined in Step 1.

•• ***multiplication & division:*** The product or quotient has as many significant figures as that of the least for any number of the problem (*i.e.*, multiplicand, divisor, dividend). Don't forget that trailing zeroes of a decimal are significant.

**COPY THIS MULTIPLICATION TABLE
TO DISTRIBUTE TO YOUR STUDENTS**

•

x	1	2	3	4	5	6	7	8	9	10
1	1	2	3	4	5	6	7	8	9	10
2	2	4	6	8	10	12	14	16	18	20
3	3	6	9	12	15	18	21	24	27	30
4	4	8	12	16	20	24	28	32	36	40
5	5	10	15	20	25	30	35	40	45	50
6	6	12	18	24	30	36	42	48	54	60
7	7	14	21	28	35	42	49	56	63	70
8	8	16	24	32	40	48	56	64	72	80
9	9	18	27	36	45	54	63	72	81	90
10	10	20	30	40	50	60	70	80	90	100

Shaded cells contain perfect squares

Table of mathematical prefixes

Prefix	Symbol	10^n	Decim[illegible]	English word
yotta	Y	10^{24}	10000[illegible]0	septillion
zetta	Z	10^{21}	10000[illegible]	sextillion
exa	E	10^{18}	10000[illegible]	quintillion
peta	P	10^{15}	10000[illegible]	quadrillion
tera	T	10^{12}	10000[illegible]	trillion
giga	G	10^{9}	10000[illegible]	billion
mega	M	10^{6}	10000[illegible]	million
kilo	k	10^{3}	1000	thousand
hecto	h	10^{2}	100	hundred
deca	da	10^{1}	10	ten
		10^{0}	1	one
deci	d	10^{-1}	0.1	tenth
centi	c	10^{-2}	0.01	hundredth
milli	m	10^{-3}	0.001	thousandth
micro	µ	10^{-6}	0.0000[illegible]	millionth
nano	n	10^{-9}	0.0000[illegible]	billionth
pico	p	10^{-12}	0.0000[illegible]	trillionth
femto	f	10^{-15}	0.0000[illegible]	quadrillionth
atto	a	10^{-18}	0.0000[illegible]	quintillionth
zepto	z	10^{-21}	0.0000[illegible]	sextillionth
yocto	y	10^{-24}	0.0000[illegible]1	septillionth

TUTOR'S FRIEND
(-17) x (-2) x (-4)
It's easy !!
Mr. Math
SIGNED NUMBERS

2. SIGNED NUMBERS

TABLE OF CONTENTS

page

●

WHAT [illegible] ER?

The figure above is called th[illegible] that the number 0 is at the center. Now, negative nu[illegible] (we say that a negative number is *less than zero*) and p[illegible] e right of 0 (we say that a positive number is *greater th*[illegible] n of a number indicates whether it is to the left or righ[illegible] at any number, positive or negative, is simply its distar[illegible]

2. SIGNED NUMBERS

Signed numbers

A *signed number* is either a positive number (*e.g.*, +3) or a negative number (*e.g.*, -3) called “negative three” or “minus three”). The “ + ” and “ - ” symbols that indicate “positive” and “negative” are *algebraic signs*, or simply *signs*. The *operators* “ + ” and “ – ” separate *terms* and indicate addition and subtraction, respectively. If a number has no sign before it, then it is positive (for example, “ 15 ” is the same as “ +15 ”), so this means *all* numbers are signed numbers. Note that any negative number, such as “ -10 ” is a positive number multiplied by -1, *e.g.*, -10 = (-1)×10.

> Example: The *expression* 5 + (-3) – 4 + 15 consists of the four *terms* 5, -3, 4, and 15 all separated by the *operators* “+,” “–,” and “+,” respectively. If you remove the parentheses from the term “(-3),” then the operator preceding that term changes from “+” to “–” and the expression becomes 5 – 3 – 4 + 15.

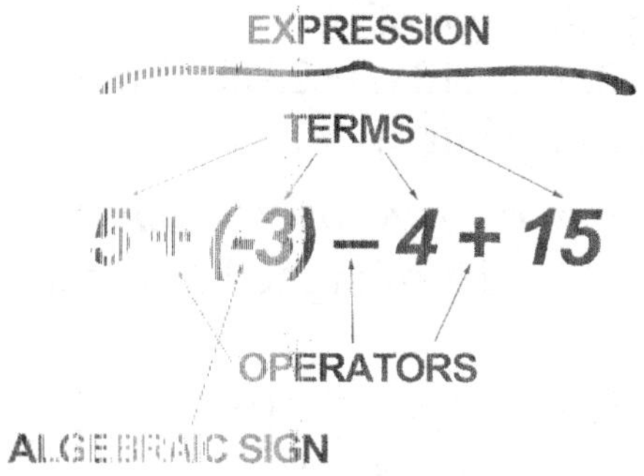

Add and subtract signed numbers

There are only two cases:

Case 1: Add (exactly) two signed numbers; either both numbers will have the same sign, or one number will be positive and the other negative.

Case 2: Add (find the sum of) more than two signed numbers.

CASE 1: ***ADD TWO SIGNED NUMBERS***

First, it will be useful to define the *absolute value* of a number: it is the number without the sign, and is usually shown as a number between two vertical lines (like this: “ | -1 | ”). Look how easy this is: the absolute value of “ +1 ” is 1 (that is, | + 1 | = 1) and the absolute value of “ -4 ” is 4 (that is, | -4 | = 4). So you just ignore the sign; that's all there is to it! Now let's look at how to add two signed numbers.

The signs of the numbers are the same (both positive or both negative) then just add them and attach the sign (positive or negative) to the sum.

> Examples: 12 + 12 = 24 and -10 – 12 = -24
> Note that “-10 – 12” is really “-10 + (-12).”

The signs of both numbers are different (one is positive, the other negative), then "take away" (subtract) the absolute value of the smaller number from the absolute value of the larger number. The sign of the answer will be the sign of the larger of the original two numbers.*

Examples: 10 – 12 = -2; -10 + 12 = 2

Subtract two signed numbers: Here, the first number (the one to subtract from) is the *minuend*; the second number (the number to be subtracted, or "taken away") is the *subtrahend.*

Example: "Subtract the second number from the first: 23, -12"
or "Subtract -12 from 23."
or "23 − (-12) = ?"

Simply *change the sign* of the subtrahend, then *add*!

i.e., to subtract -12 from 23, (*i.e.*, 23 – (-12) is same as "23 + 12 = 35"
(change the sign of -12)

To subtract 23 from 12, (*i.e.*, 12 – (+23) is same as 12 + (-23) = -11.
(change the sign of 23)

CASE 2: *ADD MORE THAN TWO SIGNED NUMBERS*

"Add," or "Find the sum of," a list of signed numbers).

Examples:

- Add (find the sum of) 19, 12, -3, 5, -10
  means 19 + 12 + (-3) + 5 + (-10) = ?
- Add (find the sum of) 14, 12, 1, 3, 6
  means 14 + 12 + 1 + 3 + 6 = ?
- Add (find the sum of) -2, -5, -10, -4, -1
  means -2 + (-5) + (-10) + (-4) + (-1) = ?

If the numbers to be added have the same sign (all + or all -) then just add them; the sum will have the same sign as the numbers added.

Examples: 14 + 12 + 1 + 3 + 6 = 36
-2 − 5 − 10 − 4 − 1 = -22

If there is a mixture of positive and negative numbers

Add all of the positive numbers to get a positive sum *then* add all of the negative numbers to get a negative sum. There are now just two signed numbers to add (this becomes Case 1).

Example: 10 + 12 − 3 + 5 − 10 = 27 – 13 = 14
the positive numbers add to 27 and the negative numbers add to -13, so we have to add "27" and "-13" 27 + (-13) = 27 – 13 = 14.

* If you find the concept of "absolute value" confusing, then just ignore the signs of the two numbers and subtract the smaller from the larger. The sign of the difference is the same as the sign of the "larger" of the original numbers.

Multiply and divide signed numbers

First, more notation: The operation of multiplication is shown in several ways. For example, "six times ten" and "three times negative two" may be expressed as "6×10" and "3×(-2)" or "6•10" and "3•(-2)" or simply "(6)(10)" and "(3)(-2)" (Notice that the negative two is shown in parentheses, and (6) and (10) have parentheses to avoid confusion, otherwise you would have "610" and "3-2," which certainly doesn't look like multiplication!) Sometimes, especially in computer programs, an asterisk is used to denote multiplication: " 6*10 " and " 3*(-2)" for example. Of course you know that when you multiply numbers, the result is the *product*.

Division may be indicated by the " ÷ " sign, or more frequently by a " — "(fraction bar or *vinculum*) or " / " (forward slash or *solidus*) or " : " (ratio).Thus," 4 ÷ 2 ", " 4/2 ", and " 4:2 " mean the same thing: "four divided by two." Yes, <u>fraction is division</u> (remember your rules for converting fractions to decimals? 1/2 = 0.5, the fraction " 1/2 " means "1 divided by 2"). The expression on top in a fraction is the *numerator*, and the expression on the bottom is the *denominator*. O.K., more on fractions in the next section; let's get back to signed numbers:the rules for multiplying and dividing signed numbers are REALLY EASY:

Multiply or divide (two numbers):

Both signs are the same (that is, <u>both</u> positive or <u>both</u> negative), then <u>multiplying</u> or <u>dividing</u> *always* produces a POSITIVE number.

The signs are different (that is, one number is positive, the other number is negative), then multiplying or dividing *always* produces a NEGATIVE number.

<u>Examples</u>: (-1) × 5 = -5; (-1) × (-5) = 5; 1 × 5 = 5
-2/5 = -0.4; -2/(-5) = 0.4; 2/5 = 0.4; 5:2 = 2.5
(Use parentheses to avoid confusion).

Multiply more than two signed numbers

The rule: If there is an <u>odd quantity</u> (1, 3, 5, ...) of negative numbers then the product of all the numbers is <u>negative</u>; otherwise, the product is positive.

<u>Examples</u>:
(-2)·(-4)·5·(-10) = -400 [there are three negative numbers, so the answer is negative.]
(-2) × (-4) × 5 × 10 = 400 [there are two negative numbers, so the answer is positive]

Raise a signed number to a positive power

A number raised to a (positive) *power* is just that number, called the *base*, multiplied by itself the number of times given by the *exponent*. Thus,

4^2 is "four-squared" and means 4 times 4 (which equals 16);
["4" is the "base" and the little "2" is the exponent (or "power")]
5^3 is "five cubed" and is 5 × 5 × 5 (which equals 125), and so on.

The preceding examples show pos [illegible] a positive power. Now, how about a negative number raise [illegible] ell, since the exponent just means "multiply the base by it [illegible] indicated by the exponent," we can use the rule that we [illegible] ing signed numbers in order to determine if the product is [illegible]

Examples:

$(-3)^3 = (-3) \times (-3) \times (-3) = -2$ [illegible] rd

$(-3)^4 = (-3) \times (-3) \times (-3) \times (-3) =$ [illegible]

Raise any real number to a neg [illegible]

First, we define the *reciprocal* of a [illegible] fraction of "1" over the number, or if the number is a fracti [illegible] inverted fraction; thus, the reciprocal of 10 is 1/10, or 0.1; t [illegible]. All we need to know is that when a number (the "base") [illegible] *power*, it means that the reciprocal of that base is raised to t [illegible] ated by the exponent.

Thus, $6^{-2} = 1/6^2 = 1/36$; and, of cou [illegible] It's as simple as that!

Taking the root of a signed nu [illegible] complicated, but still straightforward. However we wo [illegible] y math textbook will describe the rules.

NOTE: Raising a complex numb [illegible] power is beyond the scope of this book.

Opposite of a number - revisited

The *opposite* of an integer, numbe [illegible] ssion is its negative.

Examples:

The opposite of 15 is -15; the oppo [illegible] of $(x + 1)$ is $-(x + 1)$. Clearly, the sum of an integer [illegible]. A number and its opposite are the same distance [illegible] er line. A number that is divisible by another is al [illegible]'s opposite. Thus, 10 is divisible by -2 and -5 as [illegible]

The opposite of a number is also [illegible] *se*; that is, a number added to its additive inverse sums [illegible] way to move terms of an equation from one side of the e [illegible].

Example:

[illegible]

Inserting (adding) the additive in [illegible] 13 on both sides of the equation, we get

$x + 13$ [illegible]

from which we compute

$$x + 0 = 1 - 13$$

so that

$$x = -12.$$

Note that any number or term in an equation can be replaced by its opposite on the other side of the "=" symbol.

Thus $5x^3 - 2x^2 + 54x = 10$ is identical to $5x^3 = 2x^2 - 54x + 10$.

Informally, just move a term from one side of the equation to the other and change its operation sign as shown above.

THE
MATH TUTOR'S FRIEND
mixed numbers
LCD
It's easy !!
1/3 + 1/2
rationalize
Mr.
Math
FRACTIONS
2/17

3. FRACTIONS & DECIMALS

TABLE OF CONTENTS

3. FRACTIONS AND DECIMALS

Fractions

A *fraction* is a comparison of two numbers, and is expressed as one number "over" another. Examples of fractions: 1/2, 3/4, 6/9, 12/7, 5:2 and 5/1 (the last four of the list are special cases of fractions we'll be discussing later). It is important to remember that a denominator of zero is never allowed because it is *undefined* in mathematics. The "top" number is the *numerator* and the "bottom" number is the *denominator*. You may think of fractions as the ratio of two numbers (the numerator to the denominator), or as the denominator being a collection of equal parts and the numerator as the number of those parts (*e.g.*, 3/4 is same as 3:4, or 3 out of 4 equal parts).

> Example: Let's say that you have two pizzas, each sliced into 6 parts. If you take 3 slices then you have 3 out of 6 slices or 3/6 of a pizza (half a pizza). If others want pizza and you selfishly take 6 slices out of 6 parts then you have 6/6 or a whole pizza; if you are greedy and take 9 slices, you have 9 out of 6 parts or 9/6; in other words an entire pizza and 3 parts of the other.

A special kind of fraction called the *dyadic fraction* (or *dyadic rational*) is a fraction whose denominator is a power of 2; thus, 1/2, 3/4, 1/8, N/2^n are dyadic fractions. In every case - and this is important - a *fraction* represents *division*, so 3/4 and 3:4 mean "three divided by four," and 4/2 means "four divided by two." We deal here with *simple fractions,* which are expressed as one integer "over" another; examples are 2/1; 4/8; 10/10; 25/5; and 30/7. Other numbers that are not simple fractions can be converted to simple fractions; examples are

$$\frac{1}{{}^1/_3};\ \frac{{}^2/_3}{{}^1/_1};7;\ \frac{{}^3/_7}{5};\ \frac{2.5}{5};\ \frac{1}{1.2};\text{and } 0.125$$

Simple fractions are sometimes referred to as *vulgar fractions,* a term we shall see again when addressing the *Egyptian fraction.* This section deals only with the handling of simple fractions – adding, multiplying, dividing, and raising to some power. Note that ***all*** numbers (here, let's restrict the discussion to integers) are simple fractions having an "invisible denominator" of value 1.

> Examples:
> $30 = \frac{30}{1}$; $2 = \frac{2}{1}$; and $1 = \frac{1}{1}$. Knowing this fact will prove to be useful later.

There are three ways to show a negative fraction:

$$(-2)/3 = -\frac{2}{3} = 2/(-3)$$

Reducing a fraction to its lowest terms

Consider the fractions 1/2 and 4/8 and 150/300. They are *equivalent fractions*; that is, they mean the same thing because 1 divided by 2 is the same as 4 divided by 8 is the same as 150 divided by 300 which is, 1/2 or, in decimal form, 0.5.

The last two fractions (4/8 and 15[illegible]) can [illegible] to 1/2 by dividing the numerator and denominator by 4 a[illegible] other words, we have reduced 4/8 and 150/300 to *lowes*[illegible]

Add and subtract fractions

Consider the three cases:

Case 1: two or more fracti[illegible]n) denominators.

Case 2: two fractions with [illegible].

Case 3: three or more frac[illegible]minators.

NOTE: The rules for handling si[illegible] fractions as well as to integers and other numbers.

Case 1: Add fractions with equ[illegible]

Example: Add (find the sum of) [illegible]/19.

$$\frac{1}{19} + \frac{3}{19} [illegible] + \frac{7}{19} [illegible]$$

The denominators of all the frac[illegible]al; that is, there is a *common denominator* ("19"). T[illegible]mmon denominator, simply add the numerators and [illegible]mon denominator:

Case 2: Add two fractions with [illegible]rs

REMEMBER: You can add frac[illegible]) only if all fractions to be added have a common den[illegible]e, you just can't add $\frac{1}{3}$ and $\frac{1}{2}$ directly No Way!

Notice that $\frac{1}{3} = \frac{2}{6}$; as we learned e[illegible]nt fractions, and $\frac{1}{3}$ is simply the fraction $\frac{2}{6}$ reduced to i[illegible]$\frac{1}{3} = \frac{4}{12} = \frac{8}{24} = \frac{160}{480}$ (why?). Look again at adding $\frac{1}{3}$ a[illegible]$\frac{2}{6}$, and note also that $\frac{1}{2} = \frac{3}{6}$. Now we have two fracti[illegible] equivalent to the two fractions that we want to add, *i.e.*,[illegible]more, the new equivalent fractions have a common d[illegible] 1 above, we add the numerators and put the result ove[illegible]ator, 6. To summarize the example, then,

$$\frac{1}{3} + \frac{1}{2} = [illegible]$$

Cross-multiplying

HERE'S A TRICK! First, make s[illegible]receding discussion as to how to add two fractions with di[illegible]ou find a common denominator, and then simply add the [illegible] easy, but suppose you add $\frac{1}{3}$ and $\frac{3}{5}$, or $\frac{6}{17}$ and $\frac{3}{8}$, y[illegible]mon denominator, and the easiest way to do that is by the "[illegible]*ying.* Look at $\frac{1}{3} + \frac{3}{5}$. Multiply the numerator of the first [illegible]minator of the second fraction ("5"). The product, of course, is 5. Now multiply the numerator of the second fraction by the denominator of the first fraction; this yields 3×3, or 9. Now add the two products just calculated (5 + 9 = 14) and put the result over the product of the two denominators (3×5 = 15). The result is the fraction $^{14}/_{15}$. It took an entire paragraph to explain it, but the concept is easy. Cross-multiplying is just multiplying each numerator by the "other" denominator, adding (algebraically) the two products and then putting the sum over the product of the two denominators. If the result is not a fraction in lowest terms, then reduce it to lowest terms. That's all there is to it. Try the two problems below:

A. Add (find the sum of) $^1/_2$ and $^5/_8$. $^1/_2 + {^5/_8} = ?$

B. Find (the sum of) $^1/_2$ and $^1/_{17}$. $^1/_2 + {^1/_{17}} = ?$

Finally, a point to remember -- the technique of cross-multiplication only works for adding (or subtracting) two fractions. However, if you have three or more fractions, you could cross-multiply two of them then cross-multiply the result with another fraction and so on until you are left with two with which you can do a final cross-multiplication. However, there is a better way to handle the addition of three or more fractions, and that is "Case 3" below.

Case 3: Add more than two fractions with unequal denominators

Example: Add (find the sum of) 1/19, 3/11, 5/19, 7/11, 2/3.

$$\frac{1}{19}+\frac{3}{11}+\frac{5}{19}+\frac{7}{11}+\frac{2}{3}=?$$

First, combine the fractions with *like* denominators. In this example, two fractions have "19" for denominators, two have "11" for denominators, and only one with a denominator equal to "3." After combining like denominators, we have the sum of three fractions with three different denominators.

$$\frac{6}{19}+\frac{10}{11}+\frac{2}{3}$$

There are two approaches to solving the above:

a) ***Pairwise approach***: Add two of the fractions (say, using the cross-multiplication technique), then take the result and add it to the third original fraction.

b) ***Least Common Denominator approach***: Let's begin with an example; suppose that you want to add the following fractions:

$$\frac{2}{3}; \frac{1}{2}; \frac{1}{8} \text{ and } \frac{5}{30}.$$

We convert these fractions to equivalent fractions having a common denominator.

Let's take up the side issue of *factoring* an integer. *Factoring completely* is simply finding the set of prime numbers that, when multiplied together, will yield a product that is the original number; in other words, we are breaking down a number into its *prime factors*. Since the number "1" is a *factor* of any number, we need not deal with it. For example, the prime factors of the number 6 are 2 and 3, because 2 × 3 = 6; the prime factors of 8 are 2 × 2 × 2 because that product is 8. Get the idea?

Least Common Denominator

So now let's look at our fractions, specifically the denominators.

$$\frac{2}{3};\ \frac{1}{2};\ \frac{1}{8} \text{ and } \frac{5}{30}.$$

They are 3, 2, 8, 30. Their prime factors are shown below following the "=" signs:

3 = 3 (actually, 1×3, but we're ignoring the 1s,)
2 = 2
8 = 2 × 2 × 2
30 = 2 × 3 × 5

Let's build a worksheet that puts each factor in a unique column as shown below. The number to be factored is in boldface at the left, so each number has a row of factors lined up into columns of equal value. The first three columns contain the "2" factor, the fourth column the "3" and the last column the "5." Note that there is no "4" column (why not?)

2	2				
3				3	
8	2	2	2		
30	2			3	5
Factors:	2	2	2	3	5

The product of the factors is 120.

Note that we use the number from each column once because some are "common." So the least common denominator (LCD) is 120. We must now determine new equivalent fractions. See the table below. The fractions to be added are

Least Common Denominator (LCD) = 120			
Original Numerator	*Denominator*	*LCD/Denominator*	*New Numerator*
2	3	120/3 = 40	2 × 40 = 80
1	2	120/2 = 60	1 × 60 = 60
1	8	120/8 = 15	1 × 15 = 15
5	30	120/30 = 4	5 × 4 = 20

Let's summarize the process for finding the least common denominator (LCD). Having found the LCD = 120, we next convert each denominator to 120. So, for each fraction we ask "what number do we multiply the original denominator by in order to get a value of 120?" Well, the first denominator is 3, so in order to get 120, we need to multiply 3 by 40 (3×40 = 120); the number 40 is obtained by dividing the denominator into the LCD (120/3 = 40) as shown in the table. If we multiply the denominator of a fraction by 40, then we must also multiply the original numerator by the same number, so the "new" numerator is 2×40 or 80, as shown in the table above. Proceeding with the other fractions we arrive at the following results: $\frac{2\times 40}{3\times 40}$; $\frac{1\times 60}{2\times 60}$; $\frac{1\times 15}{8\times 15}$; and $\frac{5\times 4}{30\times 4}$.

The fractions to be added are $\frac{80}{120}$; $\frac{60}{120}$; $\frac{15}{120}$; and $\frac{20}{120}$.

So the problem becomes $\frac{80}{120}+\frac{60}{120}+\frac{15}{120}+\frac{20}{120}$.

And putting all over 120, the common denominator, we have

$$\frac{80+60+15+20}{120}=\frac{175}{120}=\frac{35}{24}$$

an improper fraction reduced to lowest terms.

Now, this was a long discussion, but you should be able to do most quiz problems in your head.

One rule if you get stuck: the product of all of the denominators is always a common denominator, it just may not be the *least* common denominator, and you will have to reduce the result to its lowest common terms. For example, in the preceding example we could have used the product of the denominators, namely, 3×2×8×30 = 1,440. You still would have to determine the factors of the denominators and numerators in the manner we used above. Try it!

Multiply fractions

Multiplying fractions is ***easy***: simply multiply the numerators to form the numerator of the product, then multiply the denominators to form the denominator of the product.

Example: $\frac{2}{3}\times\frac{1}{2}\times\frac{1}{8}=\frac{2\times 1\times 1}{3\times 2\times 8}=\frac{2}{48}=\frac{1}{24}$.

To multiply an integer by a fraction, note that every integer has an "invisible denominator" of one, thus:

$$15\times\frac{2}{3}=\frac{15}{1}\times\frac{2}{3}=\frac{15\times 2}{1\times 3}=\frac{30}{3}=\frac{10}{1}=10\,.$$

So, all you need to do is multiply the numerator of the fraction by the integer and use the same denominator, which, as you know, means division, as shown in the example above.

Divide fractions

Dividing fractions also is easy; [illegible] ided is the *dividend*, and the number to divide by is [illegible] divisor is a fraction, you merely *invert* (take the recip[illegible] of t[illegible] and then multiply the inverted divisor by the dividend

Example: Divide $\frac{3}{5}$ by $\frac{7}{8}$. So th[illegible] $= \frac{3}{5} \times \frac{7}{8} = \frac{24}{35}$

That's all there's to it! By the wa[illegible] a fraction as shown above is called a *complex fraction* [illegible] "complex number").

Useful facts about fractions

Reduce a fraction to its lowest te [illegible] nd denominator have a common factor, then the fracti[illegible] terms. For example, the fraction 120/175 has a comm[illegible] example, 5 is also the *greatest common factor* (GCF) o[illegible] g the numerator by 5 yields 24; dividing the denomin[illegible] the fraction 120/175 reduced to lowest terms is 24/35. [illegible] s are equivalent fractions, so their "values" (as deter[illegible] denominator into the numerator) have not changed.

All fractions are ratios: The rati[illegible] n be expressed as 2/3; another example: 4/5 is the same [illegible]

All integers are fractions/All numb [illegible] ed earlier, all numbers have an implied denominator of va[illegible] 55 = 2.55/1; π = π /1; ***x*** = *x*/1; *etc*. We used this fact earli[illegible] integer by a fraction.

All Mixed numbers are fractions: [illegible] nbinations of integers and fractions. The following shou[illegible]

$2\frac{3}{4} = 2 + \frac{3}{4} = \frac{2}{1} + \frac{3}{4} = \frac{8}{4} + \frac{3}{4} = $ [illegible] $(4 \times 2 + 3)/4 = 11/4$.

All fractions are divisions: A fract[illegible] ivision; it indicates the numerator divided by the denomin[illegible] or is the dividend and the denominator is the divisor. For [illegible] means 3 divided by 5.

All fractions are decimals: To co[illegible] imal, simply carry out the indicated division; for example [illegible] divided by 5 or 0.6, a decimal.

That's all there is to it! If the divi[illegible] t even," then carry it out to the desired number of decim[illegible] nes 0.667 rounded to three decimal places.

Canceling fractions which equal one: Any fraction for which the numerator and denominator are the same has value 1. Thus,

$$\frac{2}{2}=\frac{111}{111}=\frac{1,000}{1,000}=\frac{x}{x}=1.$$

Order a set of numbers in increasing (or decreasing) value

A problem frequently found on SAT type tests is: "***Which is larger, 3/5 or 0.66?***"

To answer this, convert the fraction to a decimal and the comparison becomes easy; thus, 3/5 (= 0.60), so 0.66 is greater than 0.60 by inspection.

You may also get a question like the following:
"Order the numbers below from largest to the smallest"

$$\frac{2}{3};\ \frac{3}{5};\ 0.45;\ \frac{10}{4};\ 0.24$$

Go ahead and do this as an exercise; just convert each fraction to a decimal and then put the decimals in decreasing order.

Rationalize the denominator of a fraction

A fractions should not have radicals in the denominator (there are exceptions, especially in statistics). If, as the result of solving a problem, you do get a radical in the denominator then you should *rationalize* the denominator.

We'll examine two cases: $\frac{3}{\sqrt{7}}$ and $\frac{6}{1+\sqrt{2}}$

In the first case, note that if you multiply the denominator by itself (square it) then you get rid of the radical sign; but if you multiply the denominator by $\sqrt{7}$ then you must multiply the numerator by that number as well (why?) Thus we have

$\frac{3 \times \sqrt{7}}{\sqrt{7} \times \sqrt{7}} = \frac{3\sqrt{7}}{7}$ and that's all there is to it!

For the second case, we introduce the *conjugate pair*. The conjugate of $x + y$ is $x - y$. In our example, the conjugate of the denominator $1+\sqrt{2}$ is $1-\sqrt{2}$. Now, you should recognize that the product of a binomial and its conjugate is the *difference of two squares*,

$$(x+y)(x-y) = x^2 - y^2.$$

Getting back to case 2: $\frac{6}{1+\sqrt{2}}$. If we multiply both numerator and denominator by the conjugate of the denominator, we get

$$\frac{6 \times (1-\sqrt{2})}{(1+\sqrt{2})(1-\sqrt{2})} = \frac{6(1-\sqrt{2})}{-1} = -6(1-\sqrt{2})$$

In both cases we have eliminated the radical sign in the denominator.

Unit fraction

A *unit fraction* is a fraction in which the numerator is 1.

Example: the fraction 5/8 can be decomposed into a sum of unit fractions:

5/8 = 1/8 + 1/8 + 1/8 + 1/8 + 1/8.

Improper fraction

An *improper fraction* is one in which the numerator is greater than the denominator; for example, 12/5 is an improper fraction.

Mixed number

A *mixed number* consists of an integer and a fraction. Thus, 2 1/5 is a mixed number.* A mixed number can be converted to an *improper fraction*:

Example: 2 1/5 = 2 + 1/5 = 10/5 + 1/5 = 11/5, an improper fraction.

Greatest common factor (GCF)

Problem: Find the greatest common factor (GCF) of 3120 and 276.

Solution: Both numbers are even, so we begin with dividing by 2, then place the quotient in the next row, to become the next dividend. Continue dividing by 2 as long as the quotient is even. In row 5, the quotient is odd, but the sum of the digits (1 + 9 + 5 = 15) is divisible by 3, so the number also is divisible by 3. The quotient in row 6 ends in 5, and because all numbers ending in 5 are divisible by 5, we use 5 as the factor.** Upon reaching 13 in row 7, we see that the only factor remaining is 13, a prime number, so we are finished. Use the same process for the number 276 (on the right of the table).

Row	*Factor*	*Number*	*Factor*	*Number*
1	***2***	3120	***2***	276
2	***2***	1560	***2***	138
3	2	780	***3***	69
4	2	390	23	23
5	***3***	195		
6	5	65		
7	13	13		

Once we determine all of the prime factors for the numbers, we search for *common factors* (*i.e.*, factors common to both numbers). For 3120 and 276 these are 2, 2, and 3 (boldfaced in the table), the product of which is 12. Therefore, the greatest common factor (GCF) of 3120 and 276 is 12, which is what we were seeking. This process can be used for any set of numbers; if no common factors are found, there is no GCF (other than 1). The table below employs the approach used for the example above. It yields the same result, namely, the GCF of 12.

* It may be useful to consider a mixed number as an integer plus a fraction; for example 14 2/3 + 5 1/3 may be rewritten as 14 + 2/3 + 5 + 1/3 = 19 + 1= 20.

**The number 195 is also divisible by 5, but we wish to proceed with the smallest prime number as factor. Note, however, that had you divided 195 by 5, the quotient would be 39, which is divisible by 3, so the end result would be the same.

Number	***Prime Factors***							
3120	2	2	2	2	3	5	13	
276	2	2			3			23
Common factors	***2***	***2***			***3***			

The table of primes is useful when determining GCF. Now suppose that 3120 and 276 are denominators of two fractions to be added, *e.g.*, 7/3120 and 5/276. We will be looking for a least common denominator (LCD), or equivalently, a *least common multiple* (LCM). Most of the work for finding the LCM is accomplished by determining the *greatest common factor* (GCF). Let's expand the table by adding a row and listing the prime factor for each column:

Number	***Prime Factors***								***Product***	***Name***
3120	2	2	2	2	3	5	13		3120	
276	2	2			3			23	276	
Common factors	***2***	***2***			***3***				12	GCF
Factors	***2***	***2***	*2*	*2*	***3***	5	13	23	71760	LCM/LCD

The two fractions can now be added:

$$\frac{7}{3120}+\frac{5}{276}=\frac{7\cdot 23}{3120\cdot 23}+\frac{5\cdot 260}{276\cdot 260}=\frac{(161+1300)}{71760}=\frac{1461}{71760}=\frac{487}{23920}$$

This problem can be solved by using a calculator. You should know the process, however, as it will be used to find GCF and LCM of algebraic expressions further on in this book series.

Find the GCF by hand

It is easy to find the factors of any integer, however large.

Example: Find all of the prime factors of 793,800. This can be done in easy steps as shown to the right.
793,800 = 2×2×2×3×3×3×3×5×5×7×7.

2 | 793800
2 | 396900
2 | 198450
3 | 99225
3 | 33075
3 | 11025
3 | 3675
5 | 1225
5 | 245
7 | 49
7

Decimals

All real numbers (numbers on the number line) can be displayed as *decimals*. Integers may show the decimal point followed by zeroes to indicate accuracy, *e.g.*, 3.00, 36.0, 1.000, *etc.*

Ratios, proportions, and odds

Ratio

A *ratio* is a comparison of one thi [illegible] ple, adults to children, or "adults/children." A ratio is sir [illegible] nple 3:2 is same as $3/2$.

Proportion

A *proportion* is an equation of tw [illegible] $\frac{3}{4} = \frac{6}{8}$ which as we have already seen are *equivalent* [illegible] ell, take the example, $\frac{3}{4} = \frac{6}{8}$; suppose one of the numb [illegible] mple, given $\frac{3}{4} = \frac{x}{8}$, we are to determine the value, *i.e.*, [illegible] ple linear equation: we multiply both sides by 8 then divid [illegible] ver. In other words, we cross-multiply as we did when add [illegible] lifferent denominators.

> Example: There are 160 adults in [illegible] adults to children is 5 to 2. How many children are th [illegible]
> Just set up the proportion by [illegible]
> 2 children/5 adults = x childre [illegible]
> Solving for x, we have $x = (2 \times 16$ [illegible] e we could solve this particular problem mentally by nc [illegible] s for each 5 children, so, given 160 adults we multiply by [illegible] because we are doing the same thing!

There is a nomenclature associat [illegible] nsider the proportion $a/b = c/d$. Here, a and d are *extr* [illegible] *means*. See the figure below. Do a cross-multiplication [illegible] = means."

> Example: Are the ratios $25/77$ an [illegible]
> Just cross multiply to see that 25 [illegible]); the product of the means is not equal to that of the [illegible] not proportional.

Mean proportional

A *mean proportional* is a proportion with equal means.

Example: th[illegible]oportion 1/3 = 3/9 (note that the means = extremes).

Example: F[illegible]e mean proportional of 3 and 75.
Note that th[illegible]n numbers are the extremes and we must determine the means: the product [illegible] means equals the product of the extremes, so we set up the equation 3×[illegible] $x = x^2$, or $225 = x^2$. Recognizing that 225 is a perfect square (15^2) we se[illegible] $x = 15$ and the mean proportional is 3/15 = 15/75.

Odds

Odds are exp[illegible]d as a ratio. For example, odds of "5 to 3 for" means that there are five [illegible]ces that the event will happen to every three chances that it will not hap[illegible] The probability of success is 62.5%. In every eight trials, there are exp[illegible]d to be five wins and three losses.

Scale factor

The *scale fac*[illegible] is defined as

Example: A[illegible] 5" photo enlarged to 9"×18" has scale factor k = (scaled value)/(orig[illegible]alue) = 9/3 = 18/6 = 3. Notice that the area 18 in^2 is scaled up to 162 i[illegible]actor of 9, or k^2. One may correctly infer that a volume is scaled-up b[illegible]

Recurring Decimals

A *decimal* may [illegible]te (3.14159), *infinite* (3.14159…), and may be *repeating* (also *recurring*) finit[illegible]2) or repeating infinite ($2.\overline{12}$). The repeating decimal $1.2\overline{34}$ may be read "one p[illegible] repeating three four," or "one point two recurring three four."

Displaying re[illegible]ing infinite decimals

Convert a no[illegible]eating finite decimal to a fraction

Example: 2[illegible] = 2,121/1,000. Informally, there are three decimal places, so remove [illegible]ecimal point and insert a "1" followed by three zeroes in the denominat[illegible]rmally, multiply numerator and denominator by 1,000; *i.e.* 2.121×(1,0[illegible]00) = 2,121/1,000.

Example: 2[illegible] = $25.25\times10^2/10^2$ = 2,525/100 = 11/4 (remember, [illegible]e multiplying by $10^2/10^2 = 1$ which doesn't change anything).

Convert infini[illegible]peating decimal to improper fraction

Steps:

(1) I[illegible]ify the repeating digit(s).

(2) L[illegible] = the repeating decimal

(3) M[illegible]ly both sides of the equation in (1) by the power of ten n[illegible]ed to move the decimal to the right of the first of th[illegible]eating digit(s).

(4) Set up another equation by multiplying both sides of the equation in (1) by a power of ten to move the decimal to the left of the repeating digit(s).

(5) Subtract the equation in Step (4) from the equation in Step (3).

(6) Solve for x, yielding the fraction equivalent of the original repeating decimal. Reduce the fraction to lowest terms.

Example :

Repeating decimal: 4.1234234...

(1) Identify the repeating digit(s). *e.g.*,
234 for 4.1234234...

(2) Let x = the repeating decimal
$x = 4.1234234...$

(3) Multiply both sides of the equation in (1) by the power of ten required to move the decimal to the right of the first of the repeating digit(s), *e.g.*, multiply by 10,000. $10000x = 41234.234...$

(4) Set up another equation by multiplying both sides of the equation in(1) by a power of ten to move the decimal to the left of the repeating digit(s), *e.g.*, multiply by 10.

$10x = 41.234234...$

(5) Subtract the equation in Step (4) from the equation in Step (3)

	$10000x = 41234.234...$	[minuend]
–	$10x = 41.234$	[subtrahend]
	$9990x = 41193$	[remainder]

(6) Solve for x, the fraction equivalent of the original repeating decimal. Reduce to lowest terms.

$x = 41193/9990 = 4577/1110$ [divide by 9]

Example:

Repeating decimal: 5.33...

(1) Identify the repeating digit(s). *e.g.*,
3 for 5.33...

(2) Let x = the repeating decimal
$x = 5.33...$

(3) Multiply both sides of the equation in (1) by the power of ten required to move the decimal to the right of the first of the repeating digit(s), *e.g.*, multiply by 10.

$10x = 53.3...$

(4) Set up another equation by multiplying both sides of the equation in(1) by a power of ten to move the decimal to the left of the repeating digit(s), *e.g.*, multiply by 1 (*i.e.*, 10^0).

$1x = x = 5.33...$ [no change]

(5) Subtract the equation in Step (4) from the equation in Step (3)

$$\begin{array}{r} 10x = 53.3... \\ -\quad x = 5.3... \\ \hline 9x = 48 \end{array}$$

(6) Solve for x, the fraction equivalent of the original repeating decimal. Reduce to lowest terms.

$x = 48/9 = 16/3$ [divide by 3]

As a check, carry out the division indicated by the fraction to get the original repeating decimal. Two more examples:

Example: ($2.\overline{12} = 70/33$)

Let $x = 2.\overline{12}$

$$\begin{array}{r} 100x = 212.\overline{12} \\ -\quad x = 2.\overline{12} \\ \hline 99x = 210 \end{array}$$

$x = 210/99 = 70/33$

Example: (37.1246... where the repeating digit is 6)

Let $x = 37.1246$

$10000x = 371246.6$	[puts the first repeating digit to left of the decimal point]
$-\ 1000x = 37124.6$	[puts the first repeating digit to the right of the decimal point]
$99000x = 3675342$	[subtract\
$x = 3675342/99000$	[solution (fraction)]
$= 612557/1650$	[lowest terms (divide by 6)]

If approximation is satisfactory then x as a mixed number is $x = 37\ 623/5000$, when reconverted to a decimal yields 37.1246, a finite decimal approximation to 37.1247, the original number rounded to four decimal places.

Managing decimals in hand computations

Addition (also addition of signed numbers): Align the decimal points in a column and do the addition, preserving the decimal points in the sum.

Example:

$$\begin{array}{r} 2.4 \\ -0.32 \\ \underline{15.} \\ 17.08 \end{array}$$

All integers are decimals

The hallmark of a decimal is the *decimal point*. It is really only a separator of the integer from a fractional part of a number, as already discussed, for example in the number 25.25, the decimal point separates the integer 25 from the "decimal" part 0.25 which we saw as the fraction 25/100 = 1/4. So, the entire number, integer and fraction, can be represented neatly by 25.25.

Multiplication. The product will [illegible] places as the sum of those in the multiplicand and mu[illegible]

Examples (showing also the san[illegible]ntific notation):
0.00254 × 0.013 = 0.00003302
(2.54 × 10 $^{-3}$ × 1.3 × 10 [illegible] 302 × 10 $^{-5}$)
15 × 0.3 = 4.5
(1.5 × 10 ^{1} × 3 × 10 $^{-1}$ [illegible])
2.1 × 0.00003 = 0.000063
(2.1 × 3 × 10 $^{-5}$ = 6.3 [illegible])

Division

- Move the decimal point of the [illegible] create an integer
- Move the decimal point in the [illegible]nber of places. Place the same number of zeroes afte[illegible]he desired number of decimal places for the quotient.
- Place the decimal point of the [illegible] that in the dividend.
- Perform the dividing operatio[illegible]

Example:

0.234) 30 = 2[illegible]

or

$\frac{30}{0.234} = \frac{30000.}{234.} =$ [illegible]05

Real-World Considerations

Some countries employ a comma i[illegible]decimal separator, and some countries use both! Consequ[illegible]notation as "25,235"or "0,25", and even 23.245.671 for [illegible]45,671. There is good reason, therefore, to employ the s[illegible]na to separate "thousands" groups in a large number (f[illegible]

Egyptian Fraction

Let N be an integer numerator an[illegible]denominator. Define the following *vulgar fractions** (a[illegible]*tion*):

Proper fraction: $N < D$; *imprope*[illegible]: A *unit fraction* has $N = 1$, and therefore is the *recipr*[illegible]*an fraction* is a finite sum of unit fractions (reciprocals [illegible]t is, of the form $1/D$; $D > 0$, such that each unit fractio[illegible]s.

Example:
The vulgar fraction ¾ can be [illegible] fraction ½ + ¼ .

* There are various other definitions for the v[illegible]

Egyptian fractions are not necessarily unique; for example,

$^4/_5$ can be expressed as $\frac{1}{2} + \frac{1}{4} + {}^1/_{20}$ and also as $\frac{1}{2} + \frac{1}{5} + {}^1/_{10}$.

Over the centuries mathematicians have developed various *greedy algorithms** to find the Egyptian fraction(s) for a given proper fraction. These may be found by searching the internet, and are a bit beyond the scope of the book. Let's apply some reasoning to solve a couple of simple cases.

Example: Consider the proper fraction $^3/_5$; we note that 3 is more than one-half of 5 so $^3/_5$ must be $\frac{1}{2}$ plus something and so we can write

$$^3/_5 = \tfrac{1}{2} + x$$

Solving for x

$$x = {}^3/_5 - \tfrac{1}{2} = ((6-5))/10 = {}^1/_{10}$$

so we have

$$^3/_5 = \tfrac{1}{2} + {}^1/_{10}$$

which is easily confirmed. The right-hand side of the equation is an Egyptian fraction equivalent to $^3/_5$.

Example:

Find an Egyptian fraction for the vulgar fraction $^8/_{11}$.

Note that 8 is more than one-half of 11 and we can write immediately

$$^8/_{11} = \tfrac{1}{2} + x\,.$$

Therefore,

$$x = {}^8/_{11} - \tfrac{1}{2} = ((16-11))/22 = {}^5/_{22}$$

and

$$^8/_{11} = \tfrac{1}{2} + {}^5/_{22}$$

now we must find the Egyptian fraction for $^5/_{22}$:

A little thought reveals that 5 is more than one-fifth of $^5/_{22}$, so

$$^5/_{22} = \tfrac{1}{5} + y$$

from which

$$y = ((25-22))/110 = {}^3/_{110}$$

and $^3/_{110}$ is easily partitioned into

$$^1/_{110} + {}^2/_{110} = {}^1/_{110} + {}^1/_{55}\,.$$

Putting all of this together yields an Egyptian fraction for $^8/_{11}$:

$$^8/_{11} = \tfrac{1}{2} + {}^1/_5 + {}^1/_{55} + {}^1/_{110}$$

which is easily verified by using a calculator to add the divisions (recall that a fraction is a division).

* Algorithms that [illegible] solutions by making local optimal decisions, choosing the best option at each step without considering the overall problem.

Summary: Egyptian Fraction

The *Egyptian fraction* is a finite sum of unit fractions (reciprocals of whole numbers) of the form $1/\boldsymbol{D}$; $\boldsymbol{D} > 0$ such that each unit fraction is unique (differs from the others).

•

THE MATH TUTOR'S FRIEND

squircle

mentally square numbers

GCF

divisibility

MULTIPLY BY
5, 25, 125
9, 99, 999

L CM

HANDY MATH TIPS

4 HANDY MATH TIPS
TABLE OF CONTENTS

4. HANDY MATH TIPS

Measurement - choose sensible units

If you describe a trip you do not say "It was a one million five-hundred eighty-four thousand-foot trip." You might say "It was a three-hundred-mile trip." In other words, you used a sensible unit of measurement. So, be reasonable when employing measurement units: feet for height, ounces for postal weight, miles for longer distances, square feet for room area measurements, and so forth.

The metric system

As already stated, the metric system is employed world-wide except for the U.S. and Liberia. Scientists everywhere use the metric system. One of the advantages of the metric system is the ease of handling extremely large and extremely small units using powers of 10. The unit of length in the metric system is the meter (abbreviated as "m"). Large distances are measured in kilometers (km). "Kilo" means one thousand* so a kilometer is 1000 meters. Small lengths may be measured in centimeters (cm) 1/100th of a meter, or millimeters (mm), 1/1000th of a meter. Weight (actually, mass) is measured in grams, kilograms, milligrams, *etc.* and volume in liters, milliliters, *etc.* One advantage of the metric system is that all measurements are expressed in a power of 10 times the basic unit, for example, 1 km = 1×10^3 meters. Compare this with the English system where one must have conversion factors like "feet to miles").

Estimating metric system dimensions

It is useful to know these basics of the metric system. Rule-of-thumb estimates are:

- a *meter* is a bit more than 3 feet (3.3 feet),
- a *kilometer* is a bit more than a half-mile (0.63 miles),
- a *centimeter* is a bit less than half-an-inch (0.39 in.), and
- a *kilogram* is just over two pounds (2.2 lbs).

Divisibility rules

(NOTE: Use 51,840 as the number to test the rules below)

Divisibilityby 2, 3, 4, 5, 6, 8, 10

An integer is...

Divisible by 2 if it ends in an even number (or 0), that is, if the *ones* digit is even (or 0). For example, -34, 2, 200, 456, and 38 are even numbers.

An integer is . . .

Divisible by 3 if the sum of its digits is also divisible by 3. Thus, we know that the number 1,567,782 is divisible by 3 because 1 + 5 + 6 + 7 + 7 + 8 + 2 = 36, which is divisible by 3, so the number itself is divisible by 3.

* In the vernacular, one might hear statements such as "It cost me three K." This, of course means three thousand dollars. Also in military jargon "A click." means 1000 meters.

Divisible by 4 if the number is ev[illegible]ields an even number. Also, if the last two digits (the [illegible]isible by 4 then so is the number.
Divisible by 5 if the units digit is [illegible].
Divisible by 6 if it is even *and* divi[illegible] 3
Divisible by 7 if 9 times the unit[illegible] the remaining digits of the [illegible] 7
Divisible by 8 if the last three dig[illegible]
Divisible by 9 if the sum of its di[illegible]
Divisible by 10 if the units digit i[illegible]

There are other tests to determin[illegible]le by 11, and so forth, but these are a bit involved and [illegible]ors, probably are not worth the effort to learn.

Problem: Show that 2520 is div[illegible] 1 to 10 inclusive.

Rapid mental calculating tips

The most important prerequisite: [illegible]ION TABLE, at least to 10×10. That table is shown on pa[illegible]ORIZE IT!

Mentally square a number *end*[illegible]

Here, we'll deal with 2-digit numbe[illegible] to handle in your head. If you are asked to square 45, for [illegible]diately come back with 2,025. Here's how: multiply the *ten*[illegible]y the next higher digit (here, 5) and "tack on" 25. Thus, 4[illegible]25 gives us 2,025. This works with larger numbers, like 1[illegible]ns become a bit more difficult with numbers of more tha[illegible] secret is knowing the multiplication table!

Mentally square a number *beg*[illegible]

Square the ones digit* and tack it [illegible]us the ones digit.

Example: Square 56
(a) Add 25 to the *ones* digit of 5[illegible]
(b) Square 6 ($6^2 = 36$)
(c) Concatenate ("tack on") (a) [illegible]36

Mentally multiply two 2-digit [illegible]gits sum to 10

Squaring a number ending in 5 is a [illegible]because the *ones* digits add to 10 (*i.e.*, 5 + 5). Take two nu[illegible]e *ones* digits add to 10 (1 + 9 = 10), so as before, take the [illegible]multiply it by the next higher digit (here, 8), so 7×8 = 56, [illegible]duct of the *units* digits (1 and 9), so 1×9 = 09, and the an[illegible]Pretty slick, hey?

* If the ones digit is 1, 2, or 3 use "01," "04," [illegible]o "tack on" two digits.

Mentally squaring a number ending in "1" (using a little algebra)

Use the *binomial expansion*: $(n + 1)^2 = n^2 + 2n + 1$; just rewrite $n^2 + 2n + 1$ as $n^2 + n + (n + 1)$.

Example: Square 71 ($n + 1$ is 71 so $n = 70$)
$71^2 = 70^2 + 70 + 71 = 4{,}900 + 141 = 5{,}041$

Example: Square 111 ($n + 1$ is 111 so $n = 110$)
$111^2 = 110^2 + 110 + 111 = 12{,}100 + 110 + 111 = 12{,}321$ (See footnote*)

Mentally squaring any 2-digit number

Here again your basic algebra knowledge is put to use. Recall that to square a binomial ... we'll illustrate: $(a + b)^2 = a^2 + 2ab + b^2$. What we did was add the square of each value, a and b, and twice the product of a and b. O.K., let's do an example and find the square of 36. As you know, 36 is really 30 + 6, so using the above procedure, square 30, which is 900, and square 6, which is 36. Now, twice the product of the two numbers is $2 \times 30 \times 6 = 360$. So we need only add 900 + 36 + 360 = 1,296. You can use another "trick" here to make the addition easier to handle in your head. Note that 36 is also 40 – 4. So use 40 instead of 36 as the middle term in the answer and add 900 + 40 + 360, which is easily determined mentally to be 900 + 400 or 1,300. Now simply subtract the 4 that we added earlier to get 1,300 – 4 = 1,296, the answer.

Be sure to practice these before trying to impress your friends!

Mentally multiplying a number by 10, 100, or 1,000

This is extremely easy, but is mentioned because the next few tips require multiplying by these numbers. Simply append ("tack on") the number of zeros (append one zero to multiply by 10, two zeros to multiply by 100, and three zeros to multiply by 1,000. This tip applies to larger numbers such as 10,000, 100,000, *etc.*).

Mentally multiplying a number by 9, 99, or 999

To multiply a number by 9 is the same as multiplying it by (10 – 1). You simply append a zero to the number and subtract the original number from it.

Example: $57 \times 9 = 57 \times (10 - 1) = 570 - 57 = 513$

Similarly, multiplying by 99 is the same as multiplying by (100 – 1).

Example: $59 \times 99 = 59 \times (100 - 1) = 5{,}900 - 59 = 5{,}841$

Use the same approach to multiply by 999, 9,999, etc…

Mentally multiplying a number by 5, 25, or 125

Multiplying by 5 is the same as multiplying by 10 and dividing by 2 (because $5 = {}^{10}/_{2}$). This is always possible because multiplying by 10 converts the original

* Note the pattern: $11^2 = 121$; $111^2 = 12321$; $1111^2 = 1234321$; $11111^2 = 123454321$.

number to an even number and according to our divisibility rules it is divisible by 2.

Example: $313 \times 5 = 3130/2 = 1{,}565$.

Similarly, multiplying by 25 is the same as multiplying by 100 and dividing by 4 because $25 = 100/4$.

Example: $568 \times 25 = 56{,}800/4 = 14{,}200$.

And multiplying by 125 is the same as multiplying by $^{1{,}000}/_{8}$.

Example: $47 \times 125 = 47{,}000/8 = 5{,}875$.

Mentally multiply two integers that differ by a small *even* number

Let a, b, and m be integers such that the two integers a and b differ by m, a small even integer; that is, $b = a + m$,

and we want to calculate the product $ab = a(a + m) = a^2 + ma$.

The expression on the right consists of the first two terms of a perfect square binomial, so we add (and subtract) $m^2/4$, the square of $m/2$, half the coefficient of a, in the second term:

$$a^2 + ma = a^2 + ma + m^2/4 - m^2/4 .$$

from which

$$a^2 + ma + m^2/4 - m^2/4 = (a + m/2)^2 - m^2/4$$

To mentally calculate $ab = a \times (a + m)$:

(1) determine the "middle number:"	$a + m/2$;
(2) square a:	a^2;
(3) square $m/2$:	$m^2/4$
(4) subtract (3) from (2) :	$a^2 - m^2/4$;
(5) $a \times b = a^2 - m^2/4$	[answer] .

The most difficult step is (2): squaring the "middle number." If you can choose the problem, select integers whose "middle number" is easy to square mentally, *e.g.*, 298 and 302 which has the "middle number" 300, easily squared to 90,000 from which you subtract $16/4 = 4$ to get the answer $298 \times 302 = 90{,}000 - 4 = 89{,}996$. Let's do some examples for small values of m: 2, 4, and 6.

Example: $14 \times 16 = (15^2) - 1 = 225 - 1 = 224$.
The middle number is 15; $15^2 = 225$; $m = 2$ and $m^2/4 = 4/4 = 1$.

Example: $13 \times 17 = (15^2) - 4 = 225 - 4 = 221$
The middle number is 15; $15^2 = 225$; $m = 4$ and $m^2/4 = 16/4 = 4$.

Example: $43 \times 49 = (46^2) - 9 = 2116 - 9 = 2{,}107$
The middle number is 46; $46^2 = 2116$; $m = 6$ and $m^2/4 = 36/4 = 9$.

Mentally multiply two 2-digit numbers whose *tens* digits are equal and *ones* digits sum to 10

Example: Multiply 73 by 77.

(a) Multiply the *tens* digit by the next higher integer ($7 \times 8 = 56$)
(b) Multiply the *ones* digits ($3 \times 7 = 21$)
(c) Concatenate (a) and (b) (5621)

Multiply an *even* number by another number: "doubling and halving"

Divide the even number by two and multiply the other number by 2. Repeat until you get to a multiplication that you can do mentally. This technique may require a scratch pad, and works best if the even number is divisible by 8 (*i.e.*, the sum of the last three digits is divisible by 8).

Example: Multiply 16 by 133.
$133 \times 16 = 266 \times 8 = 532 \times 4 = 1064 \times 2 = 2{,}128$.

Multiply a number by 2, 4, 6, 8, 16, 32, or some higher power of 2

We speak of 2^n. Simply keep doubling the product n times.

Example: $77{\times}16$ $(=77{\times}2{\times}2{\times}2{\times}2)$: $77{\times}2 = 154$; $154{\times}2 = 308$;
$308{\times}2 = 616$; $616{\times}2 = 1{,}232$.

Rapidly add a column of numbers

You probably learned to add a column of numbers from right to left; try adding from left to right!

(a) Add the 1st (left-most) column digits
assuming other columns are zero (*e.g.*, 3,456 would be 3,000).
(b) Add the 2nd column digits the same way.
(c) Continue with all other columns.
(d) Answer is the sum of (a), (b), …

Example: Add the following

3,541
2,133
434
1,267
689
21

Summing the columns from left to right:

6,000 [sum of *thousands* column]
1,800 [sum of *hundreds* column]
260 [sum of *tens* column]
25 [sum of *ones* column]

then again summing these columns from left to right,

7,000 [sum of *thousands* column]
1,000 [sum of *hundreds* column]
85 [sum of *tens* and *ones* columns]

8,085 [the total]

Practice this; you may be able to get a sum more quickly than someone using a calculator!

Checking hand calculations for [illegible]

Casting out nines

The technique of checking for e[illegible]culations by "*casting out nines*" works for whole num[illegible]tations. The principle is that every whole number can b[illegible]ot; *i.e.*, a single digit from 0 through 9, by casting out [illegible] 9.*

Example:

The digital root of 12,435,379 is 1 [illegible] 9 = 34 ► 3 + 4 = 7.
Note that if we "cast out" the 9 [illegible] (or 9 and the sums 2 + 4 + 3 and 1 + 5 + 3) the digita[illegible] 3 + 3 = 7.

Checking the four mathematic[illegible]

Addition

The digital root of the sum of dig[illegible]s is equal to the digital root of sum in the original [illegible].

2504	2 + 0	2	[illegible]	get digital root 2
8131	3 + 1	4	[illegible]	get digital root 4
2856	2 + 1	3	[illegible]	with digital root 3
3062	0 + 2	2	[illegible]	get digital root 2
16553		→ 2	[illegible]	m the sum to get
6 + 5 = 11 = 1 + 1 = 2 chec[illegible]			[illegible]	t digital root 1 + 1 = 2

The sum of the digital roots: 2 + [illegible] + 2 = [illegible]s digital root 2. ✓

Subtraction

The difference of the digital root [illegible] that of the minuend is equal to the digital root of diffe[illegible] numbers.

Cast out 6 and 3 to get digital r[illegible] 2 + 1 3

Cast out 9 and 1 and 8 to get [illegible] - 3

Subtract the digital roots to get d[illegible] ⇒ 0

for the difference. [illegible] 2 + 4 = 9 (0) check!

The digital root of the difference [illegible] ✓

Product

The digital root of the product of [illegible]tors is equal to the digital root of the product of the [illegible]

* The digital root of a number is simply the r[illegible]nber by 9.

362	2	Cast out 3 and 6 to get digital root 2
×453	3	Cast out 4 and 5 to get digital root 3 2×3 = 6
163986	6	In the product, cast out 1 and 8, 6 and 3, and 9 to get digital root 6. ✓
⇓		
6 ch[illegible]		

Division

The digital ro[illegible] the remainder + product of the digital roots of the divisor and the divid[illegible] equal to the digital root of the quotient.

36[illegible]41 ÷ 721 = 511 plus 610 remainder

⇓ ⇓

5+1+1   6+1+0

⇓ ⇓

⇒ (1 × 7) + 7 ⇒ 1+4=5

check!

In the dividen[illegible]9041) cast out 3, 6, and 9 to get digital root 4 + 1 = 5.

In the divisor [illegible] cast out 7 and 2 to get digital root 1

In the quotien[illegible] the digital root is 5 + 1 + 1 = 7

In the remain[illegible]10) the digital root is 6 + 1 + 0 = 7

Just as the di[illegible] quotient + remainder = dividend; that is

721× 511 + 6[illegible] 369041 with digital root 5

their digital r[illegible] 1 × 7) + 7 = 14 and 1 + 4 = 5. ✓

Summary

As shown abo[illegible] digital root (also *repeated digital sum*) of a whole number is the single digi[illegible] obtained by an iterative process of summing digits, on each iteration using [illegible]ult from the previous iteration to compute a digit sum. The process contin[illegible]il a single-digit number is reached, *i.e.*, the digital root. This process is spe[illegible] up by "casting out nines" before computing a digit sum. The digital root, th[illegible] simply the remainder from dividing a number by 9.

If the "casting [illegible]nes" process produces non-matching digital roots, then there is an error in [illegible]lculation. However, matching digital roots do not reveal errors of transpo[illegible] (e.g., 247 has the same digital root as 427) nor errors that are multiples of ni[illegible]

Dimension analysis

This technique, described in Book 4 of this series,* is used by scientists and engineers. It is based on the fact that if you have the same "thing" in both the numerator and denominator, the result is unity:

$3/3 = 1$; $100{,}000{,}000/1{,}000{,}000 = 1$; inches/inches = 1; liters/liters = 1; $xy/xy = 1$

Got it? Now here's a sample problem to illustrate this:

Have you ever wondered how many centuries there are in a trillion seconds? Probably not, but let's do the calculation using dimension analysis. Use scientific notation to represent one trillion seconds: $1{,}000{,}000{,}000{,}000 = 10^{12}$ *seconds*. The problem is set up as shown below; note that we begin with 10^{12} seconds, and since 1 minute is 60 seconds, place 60 *seconds* in the denominator so that it will cancel with the dimension *seconds* in the numerator. Using the same rationale, we convert minutes to hours, hours to days, days to years (using 365 days/year), *etc*., as shown. If we set the problem up correctly, after cancelling seconds, minutes, hours, days, and years, we end up with *centuries*, the dimension that we are seeking.

$$\frac{10^{12}\ \cancel{s}\quad 1\ \cancel{m}\quad 1\ \cancel{h}\quad 1\ \cancel{d}\quad 1\ \cancel{y}\quad 1\ \text{century}}{60\ \cancel{s}\quad 60\ \cancel{m}\quad 24\ \cancel{h}\quad 365\ \cancel{d}\quad 100\ \cancel{y}}$$

After cancelling dimensions that appear in both numerator and denominator as shown, the problem becomes:

$$\frac{10^{12}}{60 \times 60 \times 24 \times 365 \times 100}$$

$$= 317.1 \text{ centuries!}$$

A trillion seconds is a "loooong" time; the beginning of recorded history was only about 50 centuries ago and a trillion seconds is more than six times that!

Use the same approach to handle other dimensions such as feet, inches, miles, gallons, quarts, pints, *etc*. In the so-called "real-world" problems, dimension analysis is a neat way to check your work (did you end up with the desired dimensions?). Learn dimension analysis and use it!

Use consistent units in calculations

If you are given a problem expressed in units such as meters or liters, etc., be consistent by converting units as needed and express your answer using the appropriate units. For example, the area of a square with side length 2 feet is 4 square feet. The area of a rectangle with one side of 3 feet and the other 48 inches is found by first converting 48 inches to 4 feet and then multiplying by 3 feet to get 12 feet squared (12 square feet). Of course, we could have converted the 3 feet to 36 inches to get an answer in square inches, but why not do it the easy way?

●

* Series: "The Math Tutor's Friend."

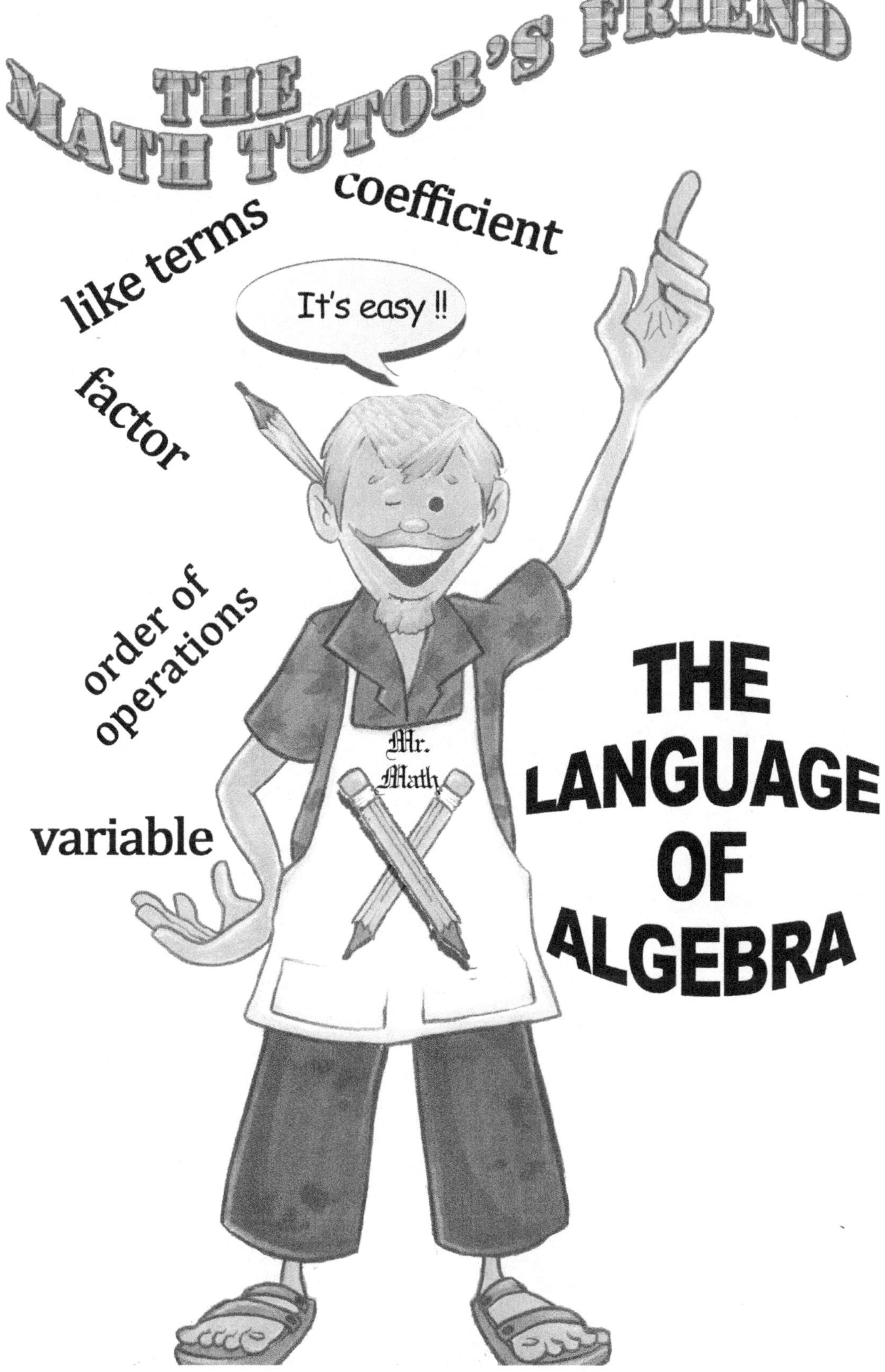
THE MATH TUTOR'S FRIEND
coefficient
like terms
It's easy !!
factor
order of operations
Mr. Math
variable
THE LANGUAGE OF ALGEBRA

5. THE LANGUAGE OF ALGEBRA

TABLE OF CONTENTS

page

5. THE LANGUAGE OF ALGEBRA

What is algebra?

Algebra is the study of mathematical symbols and the rules for their manipulation. Refer to the equation shown below. The number in parentheses shown to the right of the equation is the identifying number, so this is Equation (1), or briefly, Eqn (1):

$$x^2 - 2x + 1 = 0 \qquad (1)$$

The elements in Equation (1) separated by " – " " + " and " = " are called *terms*. As seen earlier, the " + " and " – " signs are *mathematical operators*. The *x* in the equation is a *variable*. The small " ^{2} " in the first term is an *exponent* (the *x* in that term is the *base*). The first term is "***x***-squared," (*i.e.*, $x \times x$). The number that immediately precedes a variable is a *coefficient*. The coefficient of *x* for the first term is 1 (coefficients that are 1 are not shown) and " 2 " is the coefficient of *x* for the second term. The 1 by itself on the left side is a *constant*. All the terms to the left of the " = " sign are collectively called an *expression*. There is only one term (" 0 ") to the right of the " = " sign, and it too is an expression. The entire thing (expressions and " = " sign) is an *equation*: the expression to the left of " = " is equal to the expression to the right of " = ". More on this later. By the way, we will not be solving equations in this section; we are simply learning the language of algebra. The goal is to understand and speak intelligently about topics in algebra.

> **HINT:** Most errors in solving algebraic problems are due to carelessness, like dropping a minus sign, making an error in arithmetic, or just sloppy writing (example, my cursive 2 sometimes resembles a 1, so the result, of course, could be incorrect). Keep your writing neat and don't be afraid to use paper - even if you are doing lessons or working on-line.

> **NOTE:** During many years as a volunteer tutor, I found that high school students seldom read the assigned textbook chapters before attempting homework assignments. Math books are well written these days (even though they tend to be pricey, heavy, and have too many pages and pictorial distractions). Many problems in the textbooks cite some related example in the chapter, and that example goes into detail about how to solve the particular type of problem. Unfortunately (I believe), lessons are now more and more on-line and expressed as sequences of multiple-choice questions, with comments ("great," "fantastic," "good work," *etc.*) for correct choices and explanations for incorrect choices. This may be the modern way to teach, but life does not offer multiple choice menus. Classrooms with books and human interaction should be part of education.

In my opinion, if you learned nothing else but definitions in a math course you would get at least a "C" in the course, and if you don't learn the definitions you will likely fail the course; that's how important definitions are. In fact, some math problems can be solved just by applying the applicable definition!

The language of algebra

Let's examine Equation (2) below [illegible]

$$x^3 - 3x^2 + 4x \text{ [illegible]} \qquad (2)$$

Review: In the discussions of equa[illegible] the words "expression," "term," "factor," "coefficient," "[illegible] constant," "variable," "operator" (or "operation"), "numer[illegible] and "[illegible]." There are also words like "monomial," "binomial," "trin[illegible] "po[illegible] and "like terms." If you know the meanings of these words [illegible] understand what is being discussed in algebra. More on hand[illegible] later. Each term of an expression may contain factors, va[illegible] exponents; all of these are defined below. Equation (2) consis[illegible] one on either side of the "=" sign). The expression on the [illegible] consists of five *terms* while that on the right side has o[illegible] separated by "+" and "–" signs (*operators*). Each term [illegible] contain factors, exponents, and coefficients; these are [illegible]

The terms in Equation (2) ar[illegible] $x + 1$), -5, and *ay*.

In order for the equality ("=") [illegible] hold (*i.e.*, the left side is numerically equal to the righ[illegible] one or more values of x and y that *satisfy* the equat[illegible] equation is not valid. If the equation holds (*i.e.*, is vali[illegible] then the equation is an *identity* ("≡"). We have fu[illegible] trigonometry.

A *variable* is a symbol or alpha[illegible] works as a placeholder for expression or quantities that [illegible] Equation (2) has three variables, x, y, and z. In element[illegible] are often indicated by the higher letters of the alphabe[illegible]; *w*, *x*, *y*, *z*. But this convention is not a hard-and-fast [illegible], for example, *d* could be a variable representing "dista[illegible]"

A *constant* is a fixed numerical val[illegible] = 4 or $a = 1.5$. Look at Eqn (2) and identify the constant. E[illegible] letters *a* and *c* from the beginning of the alphabet. Now, le[illegible] Equation (2). If two or more numbers or variables are mult[illegible] is a *factor* of the other. In the terms $3x^2$, $4xy/z$, $c(x+1)$, a[illegible] factors of $3x^2$. Likewise, 4, x, y and $1/z$ are factors in th[illegible] are factors in the term $c(x+1)$, and a and y are factors in [illegible]" Note that all integers consist of two or more factors (reall[illegible] 8 has factors 2 and 4, as well as 2 and 2 and 2, as well as 1 [illegible] $4 = 2{\cdot}2{\cdot}2 = 1{\cdot}8$. In fact, every integer has at least two factor[illegible]. Any positive integer that has only these two factors (*i.e.*, [illegible]mber. We learned that a coefficient is simply a numerical fac[illegible] variable. Thus, 3 is the coefficient of x^2 in $-3x^2$; 4 is the coefficient of xy/z in $4xy/z$, and the constant a is the coefficient of y in the term ay.

$$x^3 - 3x^2 + 4xy/x + c(x + 1) - 5 = ay \qquad (2)$$

Referring to Equation (2), look at the terms x^3, $3x^2$, and $4xy$. The integers 3 and 2 in small type next to variable x in the first two terms are called *exponents* (also called the *powers* of the variable x). So, x^3 has exponent 3 (or power 3, or superscript 3), and similarly for x^2. This leads to the concept of "degree."

Degree of a Variable

The term x^2 ("x-squared") means $x{\cdot}x$ and is of *degree* 2; x^3 ("*x*-cubed") means $x{\cdot}x{\cdot}x$ and is of *degree* 3; x^5 ("*x* to the fifth") is of *degree* 5; in general, x^n ("*x* to the *n*th") is of *degree n* (NOT the "*n*th degree!"). Note also that multiplying various degrees of a variable (*e.g.*, $x^3x^2x^5$) yields a product of x with exponent that is the sum of the exponents of each x (*e.g.*, $x^3x^2x^5 = x^{10}$); this is the *Law of Exponents*.

Degree of a Constant

A constant has degree zero because degrees apply to variables and a constant is just the coefficient of a variable with exponent zero, *e.g.*, $cx^0 = c$. Numbers like 5^3, 32, and c^n are constants and so have degree zero; *e.g.*, $6^2 = 6^2x^0 = 36{\times}1$ (also 5^3 and c^n have exponents, but they are *constants* and therefore of degree zero). Now that we understand the degree of variables and constants, we can proceed to the "degree of a term" and the "degree of a polynomial."

Degree of a term

As seen previously, a term may consist of any number of variables; in fact, a term may contain other terms. We will cover this later. Let's address terms with a single variable (*e.g.*, $3x^2$), and terms having two or more variables. (*e.g.*, $4xy^2z$)

Degree of a term with one variable

Examples of one-variable terms are x^5, $5y$, az, $2a^3u^4$. Note that each term contains exactly one variable (*x, y, z, and u,* respectively), Note that "*a*" is a constant so the degrees of these terms are those of the variables' exponents (5, 1, 1, and 4, respectively).

Degree of a term with two or more variables

Let's look at some examples first: $3a^4x^2y^3z$ is a term of three variables (*x, y, z*); the degree of this term is 6, the sum of the powers of the variables (remember, a^4 is a constant and is degree zero). The term $(6xy^3)/z$ contains three variables; the degree of this term is 3; it is the degree of the numerator (1 + 3) minus the degree of the denominator ("1"): 4 – 1 = 3 because $1/\mathbf{z} = 1/\mathbf{z}^1 = \mathbf{z}^{-1}$.

The language of algebra includes words like “monomial,” “binomial,” “trinomial,” and “polynomial.”

A *monomial* has a single term: $3ax^2y^3z$ is a monomial.

A *binomial* has exactly two terms; $a + b$ is a binomial.

A *trinomial* has exactly three terms; $x^2 + bx + c$ is a trinomial.

Finally, a *polynomial* is an algebraic expression of two or more terms. It should be clear that the binomials and trinomials are also polynomials, and that the terms of polynomials are separated by the operators “ + ” or “ – ”.

Degree of a polynomial

The *degree of a polynomial* is simply the degree of the term with the largest degree. A polynomial of degree 2 has the general form $ax^2 + \mathrm{b}x + \mathrm{c}$. This polynomial is a *trinomial* of degree 2 called a *quadratic*. The polynomial $ax^3 + bx^2 + cx + d$ is of degree 3. A polynomial of degree 3 is called a *cubic*.” Names are also assigned to polynomials of higher degrees.

Leading term of a polynomial

The *leading term* of a polynomial is simply the term having the highest degree. For example, the quadratic expression $ax^2 + bx + c$ has ax^2 as its leading term.

Leading coefficient of a polynomial

The *leading coefficient* of a polynomial is the coefficient of the leading term. Note that in the expression $x^2 + bx + c$, the leading coefficient is 1. A polynomial with leading coefficient of 1 is a *monic* polynomial.

Standard form for polynomials

A polynomial in *standard form* is written with the term having the highest degree followed by the remaining term(s) in order of descending degrees. The quadratic expression $x^2 + bx + c$ is written in standard form.

***Like* terms**

Like terms are terms in which the same variable is raised to the same power. The terms $4x^2$ and $2ax^2$ are *like* terms that can be combined to $2(2 + a)x^2$; note that constants are *like* terms.

Simplifying an expression (“collecting *like* terms”)

It often happens that after manipulating algebraic expressions when solving a problem that the final result contains *like terms* and the expression must be simplified. As an analogy, consider a kitchen table on which are assorted fruit such as bananas, oranges, apples, and pears. Before storing the fruit, one would consolidate them into groups of bananas, oranges, and so forth. This is the mathematical principle of collecting *like* terms. For example consider Expression (3) below:

$$2 + 4x - 3y + 2xy - 2x + 4y - 4xy + x + 2x^2y + 2xy^2 + x^2y + 6 \qquad (3)$$

Is Expression(3) a polynomial with twelve terms? No Way! Let's collect the *like* terms. There are two constants, three x terms, two y terms, two xy terms, two x^2y terms and one xy^2 term. Let's rewrite expression (3) to group *like* terms together:

$$2 + 6 + 4x - 2x + x + 4y - 3y + 2xy - 4xy + 2x^2y + x^2y + 2xy^2.$$

It's helpful to group like terms within parentheses:

$$(2 + 6) + (4x - 2x + x) + (4y - 3y) + (2xy - 4xy) + (2x^2y + x^2y) + 2xy^2$$

The like terms of Expression (3) have been collected and grouped within parentheses. (watch the signs!) The last step in collecting *like* terms is to do the math inside the parentheses and place the constant on the right:

$$3x^2y + 2xy^2 - 2xy + 3x + y + 8. \qquad (4)$$

The result of combining like terms yields an expression of five terms, as shown in Expression (4) above, which is the result of *simplifying* Expression (3).

Radicals

The radical symbol, $\sqrt{\ }$, indicates a *root*. Thus, $\sqrt{x}$, $\sqrt[3]{x}$, and $\sqrt[4]{x}$ indicate square root, cube root, and fourth root of x, respectively. The little number outside the symbol (*e.g.*, 3 and 4) is the *index* or ***root*** of the expression under the radical sign, and that *expression* is called the *radicand.* If no index is shown, then the symbol means "square root." The expression $\sqrt{a^2 + b^2}$ indicates the square root of the sum $a^2 + b^2$. The radicand of that expression is $(a^2 + b^2)$.

In general, $\sqrt[n]{x}$ indicates the nth root of the radicand x. Often it is useful to convert radical terms and expressions to terms and expressions having exponents. Thus,

$$\sqrt{x}, = x^{1/2},\ \sqrt[3]{x} = x^{1/3}, \text{ and } \sqrt[4]{x} = x^{1/4}.$$

Therefore, $\sqrt{a^2 + b^2} = (a^2 + b^2)^{1/2}$.

Function

Briefly, a *function*, $f(x)$, may be written as $y = f(\boldsymbol{x})$, and any (valid) value of $\boldsymbol{x}$ produces exactly one value of $\boldsymbol{y}$. Think of a function as a "black box" that, given an input of any value of $\boldsymbol{x}$, produces exactly one value of $\boldsymbol{y}$.

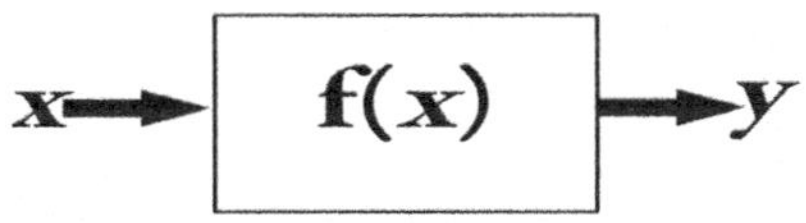

Factoring polynomial expressions

The Distributive Law

There are many "laws" in math; we deal here with the Distributive Law of multiplication over addition. Yes, addition includes subtraction: recall that 5 – 2 + 5 – 6 is the same as 5 + (-2) + 5 + (-6). We are merely adding numbers of differing signs. Thi[illegible] *algebraic addition.* Now you know why it is import[illegible]e signed numbers!

Example: $4(x + 5y)$. To apply the [illegible] multiply each term within the parentheses by the coef[illegible]

$4(x + 5y + 2z) = (4x$ [illegible] $y + 8z)$

You know, of course, that 4·5 me[illegible]lly, you "distribute" the coefficient: just multiply ea[illegible] theses by the factor outside the parentheses: $4(x - 5y$ [illegible] y

If a term is preceded by a minus [illegible], then treat that term as negative:

Example: $-4(x - 5y) = -4x - (-4$ [illegible]ul with minus signs!]

Example: $-(3x + 2y - z) = (-3x$ [illegible] the minus signs]

An expression in parentheses pr[illegible]n has "-1" as a factor which, when distributed across t[illegible]ntheses, changes the sign of each term as shown abo[illegible]ient itself may be an expression (enclosed in parenthe[illegible]he expression below, the "$(a + b)$" is a coefficient and [illegible] the terms in $(a - b)$:

$$(a + b)(a - b) = [(a + b)a - (\ldots [illegible] \ldots b - b^2) = a^2 - b^2.$$

Brackets ("[ ]") are used often to av[illegible]g., [(a + b) – (c + d)].

Other examples showing [illegible]ibutive Law:

$a(a + b + c + 2) = a{\cdot}a +$ [illegible] $+ ac + 2a$

$x^2(2x^2 - 5x + 1) = (2x^2$ [illegible] $- 5x^3 + x^2$

It is important that you complet[illegible]ributive Law before proceeding to factoring.

Factoring an expression

Factoring is simply the reverse [illegible]w. It is employed to solve and simplify algebraic expr[illegible]action always means division.

Before we discuss the mechanics [illegible]ollowing examples:

- $a/a = 1$

  [*a* divided by *a* equals 1, $a \neq$

- feet/feet = gram/gram = 1

  [any dimension divided by its [illegible]d in *dimension analysis*]

- 23,456/23,456 = 1

  [any number except zero divi[illegible]rd

- $(x^2 - 3x + 5) \div (x^2 - 3x + 5) =$

  [any algebraic expression (not ev[illegible] itself equals 1]

Why is this im[illegible]n? If you multiply or divide any expression or variable by 1, nothing is cha[illegible]! This is the "trick" of factoring.

Note that an e[illegible]ssion such as $3ab/a$ ($a,b \neq 0$) can be rearranged to $3(a/a)b$ which is equi[illegible]t to $3b$ because the *a*s "cancel" (*i.e.*, $a/a = 1$).

Let's work wi[illegible] few examples (factors were mentioned earlier in this book): We saw earli[illegible] any integer is the product of its factors. Recall this?

$$8 = 2 \cdot 4 = 2 \cdot 2 \cdot 2 = 1 \cdot 8 \text{ and } 5 = 1 \cdot 5$$

Examine Exp[illegible]on (5), repeated below:

$$ab - ax + 2a + a^2y \qquad (5)$$

Note the each [illegible] in the expression includes *a*. Enclose the entire expression in parenthese[illegible]y^2] and multiply it by $\frac{a}{a}$.

$$\frac{a}{a}(ab - ax + 2a + a^2y)$$

Change $\frac{a}{a}$ to a[illegible]hich is still equivalent to 1.

$$a\frac{1}{a}(ab - ax + 2a + a^2y)$$

Now distribut[illegible] across the terms of the expression (this leaves *a* outside the parentheses a[illegible]ctor of the expression):

$$a\left(\frac{1}{a}ab - \frac{1}{a}ax + \frac{1}{a}2a + \frac{1}{a}a^2y\right)$$

"Canceling" t[illegible]s (because $a/a = 1$) results in the expression shown below, which is a fac[illegible] (simplified) version of Expression (5).

$$a(b - x + 2 + ay) \qquad (6)$$

You can chec[illegible]ression (6) by using the Distributive Law to distribute the *a* coefficient t[illegible] the original Expression (5). Try it!

Informally, fa[illegible]ng an algebraic expression is done by

1) findi[illegible] variable, constant, term, or expression that is a coefficient (*i.e.*, "[illegible]r") of each term in the original expression, and

2) "pull[illegible]out" this common coefficient (*i.e.*, factoring) to yield the result.

Note: the fac[illegible] "pulled out" *is a factor of each and every term* in the original expr[illegible]on. The reason this is so important is that in all of algebra and subseque[illegible]ath, you will be using the factoring technique over and over again w[illegible]olving problems.

Factor a coefficient from a term that does not contain that coefficient

It is possible to factor a constant or variable out of a term that does not contain that constant or variable.

Example:

"Factor $5a$ from $3ax + 25$." Here's how, using our knowledge that dividing a term by itself equals "1:"

First, multiply the expression by $5a/5a$ (a not equal to zero), which is just multiplying by "1," so doesn't change anything. But look!

$$5a/5a(3ax + 25) = [(15a^2x/5a) + (125a/5a)] \text{ by the Distributive Law.}$$

Now we can factor $5a$ from ***that*** expression:

$$[(15a^2x/5a) + (125a/5a)] = 5a(3ax/5a + 25/5a) = 5a(3x/5 + 5/a)$$

which is easily verified.

The Number Line - revisited

Recall from the definition of integers that they may have negative or positive values or may be 0 (zero). Integers can be represented on a ***number line***, shown below.

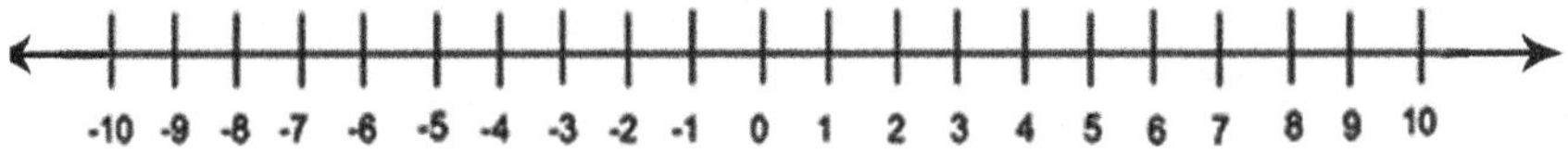

Note that zero is at the center of the line, negative integers are to the left of zero and positive integers lie to the right of zero. If one integer is smaller than another, then it lies to the left of the other integer. For example, clearly, $3 < 5$, and this can be seen on the number line. Note also from inspecting the number line that $-2 < 0$ and $-7 < -5$. The number line will help you to understand the concept of *absolute values*.

Inequalities

We have been discussing equations, *i.e.*, equalities, up to this point. Mathematics must be able to deal also with variables and expressions that are not equal. For example, 3 is certainly not equal to 5, so we may write $3 \neq 5$; also, 3 is less than 5; in fact, 3 is strictly less than 5, so we express this mathematically as $3 < 5$ (or $5 > 3$). These are called *strict inequalities*, or sometimes *strong inequalities*. Note that the arrowhead points to the smaller of the two quantities. Thus, if a is less (smaller) than b we write "$a < b$." The symbols " < " and " > " are on most computer keyboards. In some computer programs, " <> " signifies "not equal to," which is the same as " $\neq$ ". This introduction should allow you to understand the other types of inequalities shown on the next page.

Math symbols and corresponding <Alt> codes

$\boldsymbol{a = b}$	***a*** is equal to ***b***
$\boldsymbol{a \neq b}$	***a*** is not equal to ***b*** <Alt> 8800 (may not work in some apps)
$\boldsymbol{a < b}$	***a*** is less than ***b*** (*strict inequality*)
$\boldsymbol{a > b}$	***a*** is greater than ***b*** (*strict inequality*)
$\boldsymbol{a \geq b}$	***a*** is greater than or equal to ***b*** (*weak inequality*) <Alt> 242
$\boldsymbol{a \leq b}$	***a*** is less than or equal to ***b*** (*weak inequality*) <Alt> 243
$\boldsymbol{a << b}$	***a*** is much less than ***b***
$\boldsymbol{a >> b}$	***a*** is much greater than ***b***
$\boldsymbol{a \approx b}$	***a*** is approximately equal to ***b*** <Alt 247>
$\boldsymbol{a \equiv b}$	***a*** is identical to ***b*** (if *a* changes, then *b* also changes). <Alt 240>

(Contrast this with $a = b$, where if *a* changes, then the equality is no longer true.)

$\boldsymbol{a \stackrel{?}{=} b}$	"is ***a*** equal to ***b***?" (equality is unproved or yet to be proved). <Alt 8799>
$\boldsymbol{a \stackrel{?}{<} b}$	"is ***a*** less than ***b***?" (inequality is unproved or yet to be proved). <Alt 10875>
$\boldsymbol{a \stackrel{?}{>} b}$	"is ***a*** greater than ***b***?" (the inequality is unproved or yet to be proved). <Alt 10876>

NOTES:

1. ***a*** and ***b*** may be real numbers, constants, expressions, or functions.

2. Some "Alt" codes don't work in some apps or typographic fonts. If a code doesn't work in the Times New Roman font, for example, try the Cambria Math font.

3. Larger (*e.g.*, 4- and 5-digit) "Alt" codes may not work in some apps.

●

INDEX

TEST QUESTIONS

Pre-Algebra

QUIZZES

• The real number 3 is a

A. Natural number
B. Whole number
C. Integer
D. All of the above
E. None of the above.

• • Answer D. The real number 3 is a natural number, a whole number, and an integer. Note that the number 3.5 would be "None of the above."

• ♥ •

• Cut a cake into eight pieces of equal size (is that possible?) You eat three of the pieces. What fraction of the cake have you eaten?

A. 3/8
B. 8/3
C. 8/8
D. 3/3
E. None of the above.

• • Answer A. You ate three parts of eight; the applicable fraction is 3/8. (Note: answers C and D are equivalent, *i.e.*, the whole cake!)

• Are the numbers 148.5, 16.7, a[illegible] 2344 rat[illegible]

A. 148.5 and 2344 are rati[illegible] 16.7 i[illegible]

B. 16.7 and 2344 are ratio[illegible] 148.5 i[illegible]

C. Only 2344 is rational

D. All three numbers are ra[illegible]

E. None of the above

• • Answer D. The number 148.5 [illegible] 297/2 and is thus a rational number; 16.7 can be expre[illegible] 16[illegible] 2344 can be expressed as 2344/1.

• Express a comparison of the nu[illegible] inequality.

A. 23 is greater than -5

B. 23 is greater than or eq[illegible]

C. -5 is less than or equal [illegible]

D. All of the above

E. None of the above

• • Answer D. All of the above. E[illegible] are clearly not equal, it is correct to state B and C. Thu[illegible] 0), the variable *x* may take on any negative value or ma[illegible]

Here are some less-used compari[illegible]

$a << b$ *a* is much less [illegible] is much greater than *b*

$a \approx b$ *a* is approxim[illegible]

$a \equiv b$ … a is identical to b (this means that if a changes, then b also changes. Contrast this with $a = b$, where if a changes, then the equality is no longer true)

• Round the decimal 0.5555 to two decimal places

A. 0.56

B. 0.54

C. 0.55

D. 0.555

E. None of the above

• • Answer A.

The rationale is 0.5555 rounds up to 0.556 which rounds up to 0.56

• ♥ •

• Mark the locations of -3 , 0 , 0.245 , ¾ , 1½ , 2 on the number line (you may have to interpolate).

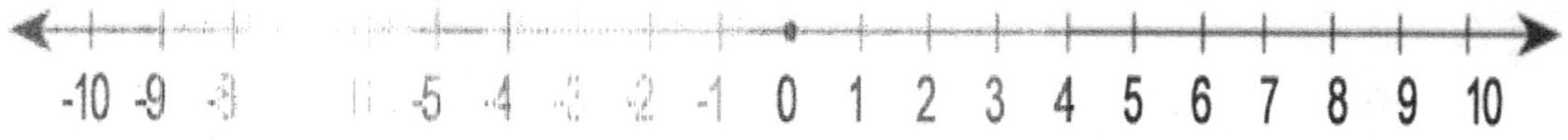

• • Answer:

Note that each tick mark is an integer; locating numbers between integers requires us to interpolate (*i.e.*, guess) or change to a more detailed number line, say with tick marks between each integer.

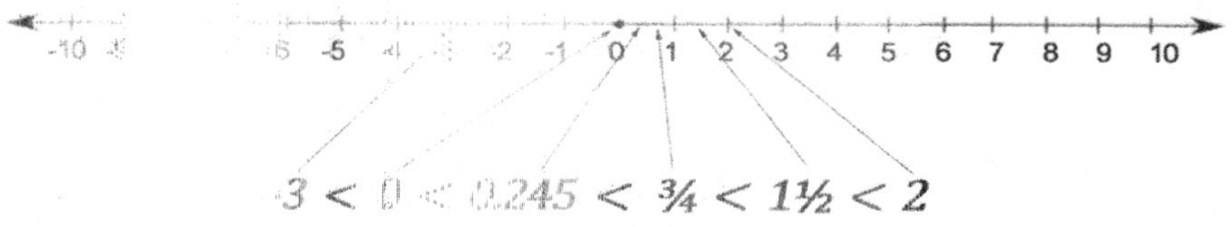

• Consider the two fractions 3/4 and 5/8. Which fraction lies to the right of the other on the number line?

A. 3/4
B. 5/8
C. They are at the same point on the number line
D. It is not possible to compare these fractions
E. None of the above

• • Answer A. 3/4. Set the fraction in A next (to the left of) to B and cross multiply to get 24 (on the left) and 20. Clearly 24 > 20 and so 3/4 is to the right of 5/8 on the number line.

Using notation: $\frac{3}{4} \overset{?}{<} \frac{5}{8}$ then cross-multiply and compare as described above.

Note that you could have also converted 3/4 to 6/8, which is clearly greater than 5/8, to arrive at the same answer. And you could have converted each fraction to a decimal number and then make the comarison.

• ♥ •

• The sum 19 + 12 + (-3) + 5 + (-10) is

A. -25
B. 36
C. 23
D. -36
E. None of the above

• • Answer C.

$$19 + 12 + (-3) + 5 + (-10) = 36 - 13 = 23$$

• The “tens” digit and the “hundredths” digit of 342.23 are, respectively,

A. 4 and 3
B. 2 and 2
C. 3 and 2
D. 6 and 5
E. None of the above

• • Answer A. The number 342 can be expressed as $3\times100+4\times10+2\times1$, from which is seen that the “3” is the “hundreds” digit, “4” the “tens” digit, and “2” the “ones” digit. Similarly, the decimal part of the number can be expressed as $2\times1/10+3\times1/100$

• ♥ •

• The length and width of a note pad would most likely be expressed in

A. meters
B. millimeters
C. centimeters
D. kilometers
E. The metric system does not apply to smaller dimensions.

• • Answer C

• Divide 12 by 5/6 (*i.e.*, 12/(⅚) [illegible]

A. 12/6
B. 5/72
C. 10
D. 72/5
E. None of the above

• • Answer D. Invert the divisor a [illegible]

12/(5/6) [illegible]

• The additive inverse and multip [illegible] e, respectively,

A. $-x^2$ and $-1/x^2$
B. $-1/x^2$ and $1/x^2$
C. x and $-x$
D. both $-1/x^2$
E. None of the above

• • Answer E.

The additive inverse of x^2 is its neg [illegible] $(-x^2) = 0$.

The multiplicative inverse of x^2 i [illegible] at is, $x^2 \times 1/x^2 = 1$.

• The multiplicative inverse of 0.5 is

A. -0.5
B. 1/2
C. 2
D. 0.5/1
E. None of the above

• • Answer C. The product of a number and its multiplicative inverse equals "1."
So $0.5 \times 2 = 1$ and "2" is the multiplicative inverse of 0.5.
Also note that $0.5 = ½$ and $2 \times ½ = 1$.

• ♥ •

• The product of $(-137) \times 45 \times (-966) \times (-33) \times 789 \times (-57)$

A. is a negative number
B. is a positive number
C. is less than one
D. is an imaginary number
E. None of the above

• • Answer B. The product of these factors is a positive number because there are an even number (4) of negative factors.

• 6/37 – 1/8 =

A. 11/296
B. 7/44
C. 5/37
D. 7/8
E. None of the above

• • Answer A.
Use cross-multiplication:

$$(8\times6 - 37\times1)/(8\times37) = (48 - 37)/296 = 11/296$$

• ♥ •

• The number "three million" would be expressed in the metric system as

A. 3×10^7
B. 3.0×10^7
C. 3,000,000
D. 3 000 000
E. None of the above

• • Answer D. A space is used instead of the comma in the metric system.
Note: Scientific notation is used in the metric system and Answer A or B would have been correct had the exponent been 6 instead of 7 as shown.

• Round 567,345,449 to the nearest thousand

A. 567,345,400
B. 567,346,000
C. 567,345,000
D. 567,000,000
E. None of the above

• • Answer C.
The rounding digit is "5," the thousands digit and the adjacent digit to its right is below 5, so the rounding digit remains unchanged. The numbers to the right of the rounding digit are replaced by "0's."

• ♥ •

• Equivalent fractions

A. Have the same denominator
B. Have the same numerator
C. Reduce to the same lowest terms
D. Cannot be multiplied together
E. None of the above

• • Answer C.
They are actually the same fraction: for example, ½, 20/40, and 500/1000 are equivalent fractions.

- The mixed number 7½ is equivalent to

  A. 15/2
  B. 7 + ½
  C. 21/3 + ½
  D. All of the above
  E. None of the above

- • Answer D

- Which symbol, if placed in the [illegible] will [illegible] statement?

  -7 [illegible]

  A. =
  B. ≤
  C. ≥
  D. Any of the above
  E. None of the above

- • Answer D. These are equivalent [illegible] the fraction on the right to get -7/9 which is equal to that [illegible] left.

• $|-34| = ?$

A. -34
B. -3.4
C. 3.4
D. 34
E. None of the above

• • Answer D

• ♥ •

• You solve a "real-world" math problem and your answer, before simplifying, is

$$\frac{\text{mile} \times 60 \text{ seconds} \times 2 \text{ minutes} \times 300 \text{ feet}}{120 \text{ seconds} \times \text{minute} \times 5280 \text{ feet} \times \text{hour}}$$

When simplified, the dimensions of the answer are

A. miles/feet
B. seconds/hour
C. miles/hour
D. feet/second
E. None of the above

• • Answer C. Other dimensions cancel out (*e.g.*, feet/feet = 1)

• The area of a rectangle is calculated by multiplying its length times its width. What is the area of a plant box measuring 8 inches by 4 feet?

A. 32 square inches
B. 32 square feet
C. 382 square feet
D. 382 square inches
E. None of the above

• • Answer D. The dimension in feet must be converted to inches before computing the area. Alternatively, the dimensions may be converted to feet: thus, 8/12 ft × 4 feet equals 8/3 square feet.

• ♥ •

• Express the integer 7 in scientific notation

A. 7^7
B. 7×10^0
C. 7×10^1
D. 7×10^{-1}
E. None of the above

• • Answer B.

$$7 \times 10^0 = 7 \times 1 = 7.$$

Any number or expression raised to the zero power equals 1.

• Find the least common multiple (LCM) of 230 and 125

A. 5
B. 230
C. 125
D. 5,750
E. 28,750

• • Answer D. The factors or 230 are 2, 5, and 23
The factors of 125 are 5, 5, and 5.

230:	2	5			23
125		5	5	5	
	2	5	5	5	23 product: 5,750

• ♥ •

• The prime factors of 1,248 are

A. 2^5, 3, 13
B. 2, 4, 8
C. 3, 7, 19
D. 1, 2
E. None of the above

• •

Answer A.

$$2 \cdot 2 \cdot 2 \cdot 2 \cdot 2 \cdot 3 \cdot 13 = 2^5 \cdot 13 = 1{,}248$$

- Rationalize $\frac{1}{2 - \sqrt{5}}$ [illegible]

  A. 2

  B. 5

  C. $2 - \sqrt{5}$

  D. $-(2 + \sqrt{5})$

  E. None of the above

- • Answer D.

$$\frac{1}{2 - \sqrt{5}} = \frac{1}{2 - \sqrt{5}} \text{ [illegible]}$$

[illegible]

- $(2/3) \div (3/4) =$

  A. 1/3

  B. 4/9

  C. 4/3

  D. 3/2

  E. None of the above

- • Answer E.

$(2/3) \div (3/4) =$ [illegible]

• 2/5 + 6/7 − 1/35 =

A. 9/7

B. 5/2

C. 8/35

D. 35/8

E. None of the above

• • Answer A.

2/5 + 6/7 + 1/35 = (14 + 30)/35 + 1/35 = 45/35 = 9/7

• ♥ •

• The reciprocal of 1.23 is

A. [illegible]

B. [illegible]

C. [illegible]

D. [illegible]

E. None of the above

• • Answer D.

1.23 = 123/100; the reciprocal is 100/123

• A light-year is approximately ___________ miles.

(Use the following conversions:

speed of light is approximately 3.00×10^5 kilometers/sec;
5 miles $\approx$ 8 km; 1 year $\approx 3.2 \times 10^7$ seconds)

A. 5 000 000

B. 5.00×10^6

C. 3.00×10^5

D. 5.85×10^{12}

E. None of the above

• Answer D.

Use dimension analysis:

(1 year) × (5 mi/8 km) × (3×10^5 km/sec) × (3.12×10^7 sec/yr)

The "year," "km" and "sec" cancel leaving 5.85×10^{12} miles

• ♥ •

• To mentally multiply 87 by 99 you would

A. "tack on" (append) 99 to 87

B. append "00" and subtract 99

C. append "00" and subtract 87

D. append 87 to 99

E. None of the above

• • Answer C.

$$87 \times 99 = 87 \times (100 - 1) = 8700 - 87 = 8613$$

• Mathematical operators

A. Include the infinity symbol (∞)

B. Separate terms in a mathematical expression

C. Are signed numbers

D. Are prime numbers

E. None of the above

• • Answer B.

The four common operators are the symbols $+$. $-$, $\times$, $\div$. Other actions such as exponentiation and taking roots are sometime called operations.

• ♥ •

• The subscripted variable x_2 signifies

A. The square of x

B. The square root of x

C. That it is a variable different from other subscripted x -variables

D. The product of x and 2

E. None of the above

• • Answer C.

The subscript is merely a label; it is sometimes used to signify the base of a number, e.g., $10_2 = 2_{10}$; that is, $10_2 = 0\times 2^0 + 1\times 2^1 = 0 + 2_{10} = 2_{10}$.

• In the expression $2 : x = 3 : 18$, the variable x

A. Equal to 3

B. Called an *extreme* of the proportion

C. Called a *mean* of the proportion

D. Is undefined

E. None of the above

• • Answer C.

• The number 39,702 is

A. Divisible by 3

B. A whole number

C. An integer

D. All of the above

E. None of the above

• • Answer D.

For "A," The sum of the digits, 21, is divisible by 3, so the number itself is divisible by 3.

• Let $x = 3$ and [illegible]; then $y = 3$. This is an example of the

A. Distributive Law

B. Associative property

C. Symmetric property

D. Reflexive property

E. None of the above

• • Answer E.

This is an example of the transitive property.

• ♥ •

• The equation [illegible] y/y where $y > 0$ is

A. Undefined

B. An identity

C. True [illegible] if $x = y$

D. True [illegible] if y is Real

E. None of the above

• Answer B.

Both sides of the equation evaluate to 1 whatever the value (real or complex) of y.

• The infinite recurring decimal 1.666... can be written as

A. 5/3

B. 1 2/3

C. 1.667

D. Any of the above

E. None of the above

• • Answer D.

Let $x = 1.666...$
then $10x = 16.666...$
$10x - x = 15$
$9x = 15$
$x = 15/9 = 5/3 = 1\ 2/3 = 1.666... = 1.667$ rounded to three decimal places.

• ♥ •

• A transcendental number can be displayed on the number line

A. TRUE

B. FALSE

• • Answer A. TRUE.

A transcendental number is a real number. Any real number can be displayed on the number line.

• The value of $(-1)^{-17}$ is

A. -27

B. 27

C. -1

D. 1

E. None of the above

• • Answer C.

The negative exponent means that the base is a reciprocal. We have $1/(-1)^{17}$. The exponent is odd, and -1 raised to any odd exponent remains as -1.

• ♥ •

• The integer 2520 is divisible by

A. 1, 2, 3, 4
B. 6, 8, 10
C. 5, 7, 9, 15, 21, 35
D. All of the above
E. None of the above

• • Answer D.

You can confirm this by dividing by each listed integer or (more easily) by finding the prime factors of 2520: 2, 2, 2, 3, 3, 5, 7 and noting that $2^2 = 4$,
$2^3 = 8$,
$2 \times 3 = 6$, $3^2 = 9$, $2 \times 5 = 10$, $3 \times 5 = 15$, $3 \times 7 = 21$, and $5 \times 7 = 35$.

• The Eiffel Tower in Paris rises t[illegible]ately. A vendor at the foot of the tower sells miniat[illegible]1. What is the scale factor?

A. 2544

B. 1/212

C. 212

D. 1/2544

E. None of the above

• • Answer D.

First, put both measurements int[illegible]5" = 5/12 feet.

The scale factor, k, is k = (scaled[illegible]) = 5/(12×1060) = 5/12,720 = 1/2,544.

• A six-inch tall "Washington Mo[illegible]is scaled at 1/1,100.

How tall is the Monument?

A. 1100 feet

B. 110 feet

C. 550 feet

D. 5,555 feet

E. None of the above

• • Answer C.

Original height = Scaled height/s[illegible]12"× 1/1100) = 1100/2 = 550 feet.

• A 2 by 1 (L× [illegible] foot photo print is expanded to a 5,000 square foot sign. What are the [illegible] sions of the sign (L×W)?

A. 10,0 [illegible] et by 5,000 feet

B. 100 [illegible] by 50 feet

C. 1,0 [illegible] et by 500 feet

D. 20 [illegible] y 10 feet

E. Non [illegible] he above

• • Answer B.

The area of th [illegible] ginal print is 2 square feet; the area of the sign is 5,000 square feet. Th [illegible] fore

k^2 = (s [illegible] l area)/(original area) = 5,000/2 = 2,500

The sca [illegible] ctor is the square root of 2500, so $k = 50$

Applying the [illegible] ion of scale factor to find the length (L) and (W) of the sign:

k = (sca [illegible] imension)/(original dimension)

$k = 50$ [illegible] 1’ = L/2’ from which W = 50’ and L = 2× 50’ = 100’

THE MATH TUTOR'S FRIEND

INFORMAL REVIEW OF HIGH-SCHOOL MATH TOPICS FOR TUTORS

- 1 -

' Pre-Algebra '

Made in the USA
Monee, IL
05 November 2026

67382707R00066